# Studies in Computational Intelligence

Volume 1269

The series "Studies in Computational Intelligence" (SCI) publishes new developments and advances in the various areas of computational intelligence—quickly and with a high quality. The intent is to cover the theory, applications, and design methods of computational intelligence, as embedded in the fields of engineering, computer science, physics and life sciences, as well as the methodologies behind them. The series contains monographs, lecture notes and edited volumes in computational intelligence spanning the areas of neural networks, connectionist systems, genetic algorithms, evolutionary computation, artificial intelligence, cellular automata, self-organizing systems, soft computing, fuzzy systems, and hybrid intelligent systems. Of particular value to both the contributors and the readership are the short publication timeframe and the world-wide distribution, which enable both wide and rapid dissemination of research output.

Indexed by SCOPUS, DBLP, WTI AG (Switzerland), zbMATH, SCImago.

All books published in the series are submitted for consideration in Web of Science.

Hung Phu Bui · Le Pham Hoai Huong
Editors

# Artificial Intelligence in Education and Academic Research

## Opportunities, Challenges, and Ethical Issues

Springer

*Editors*
Hung Phu Bui
Faculty of Foreign Languages
Ton Duc Thang University
Ho Chi Minh City, Vietnam

Le Pham Hoai Huong
Hue University of Foreign Languages
and International Studies, Hue University
Hue, Vietnam

ISSN 1860-949X ISSN 1860-9503 (electronic)
Studies in Computational Intelligence
ISBN 978-3-032-18391-0 ISBN 978-3-032-18392-7 (eBook)
https://doi.org/10.1007/978-3-032-18392-7



This Springer imprint is published by the registered company Springer Nature Switzerland AG
The registered company address is: Gewerbestrasse 11, 6330 Cham, Switzerland

# Preface

Artificial Intelligence (AI) has become an integrative and transformative tool in education and academic research, redefining how knowledge across disciplines is constructed, delivered, and assessed. The ubiquitous use of such tools as ChatGPT, NotebookLM, and other intelligent systems has not only expanded opportunities in personalized and data-driven learning but also introduced challenges related to integrity, equitable access, and teacher AI literacy. Within this dynamic context, understanding the affordances and constraints of AI in education and academic research has proved to be an urgent endeavour.

This edited volume *Artificial Intelligence in Education and Academic Research* presents empirical and conceptual manuscripts across contexts and disciplines, offering insights into how AI has been applied in teaching, learning, and research practices. Part One, Conceptual and Policy Foundations of AI in Education and Research, introduces the foundations and policy for AI integration with Chapter "Generative Artificial Intelligence (GenAI) in Language Education: Current Trends and Perspectives" which discloses findings related to generative AI trends in language education. Chapter "AI Integration into Higher Education Curricula: A Multiple-Case Study of Key Universities in Vietnam" in this part presents a multiple-case study of how leading universities in Vietnam have incorporated AI into higher education curricula. Part Two, AI Applications in Teaching and Learning Contexts, presents findings related to AI use in classroom settings. For example, Chapter "Turkish Students' Use of NotebookLM in an Online Digital Literacy Course" reports how NotebookLM is integrated in digital literacy courses for Turkish students or how chatbot promotes EFL teachers' confidence and technology acceptance. Following Chapter "Impacts of Chatbot Training on EFL Teachers' Confidence and Acceptance of AI Technologies" reports empirical findings about the effects of training on Vietnamese EFL teachers' confidence and acceptance of AI technologies, and Chapter "Building Audiobooks with AI Support as Digital Resources for ESL Learning" in this part gives details of AI-generated audiobooks as learning resources. Part Three, *AI-Mediated Learning and Interaction* comprises four chapters. Chapter "Chatbot-Human Interaction in Vietnamese EFL Speaking Classes: Exploring the Unchartered Territory" analyzes human–chatbot interactions in EFL speaking classes, followed by Chapter "EFL Learners' Engagement in ChatGPT–Generated Speaking Tasks from a Text-Driven Approach: A Case in Vietnam" exploring learner engagement with ChatGPT-generated speaking tasks. Chapter "Artificial Intelligence-Supported Willingness to Communicate and Intercultural Communication" argues for the benefit of using AI to improve willingness to communicate across cultures. This part also includes Chapter "Artificial Intelligence–Enhanced Scaffolding and Peer Collaboration in Culturally and Linguistically Diverse Classrooms: Evidence from Asia" on AI-enhanced scaffolding and peer collaboration in multilingual Asian classrooms. As the last part of the book, Part Four focuses on *AI Feedback and Learning Assessment*. Chapter "AI-Mediated Feedback in L2 Writing: Characteristics of Prompts, AI Tools, and Learners' Feedback Use" outlines the main interests in AI-mediated written feedback and argues for how learners should use AI feedback to improve their writing. Chapter "Ethical, Pedagogical, and Research Implications of AI Use in Education and Academic Research" concludes this edited book, presenting ethical, pedagogical, and research implications for AI integration into education and academic research.

The landscape of AI integration in education and research is evolving rapidly; thus, it is argued that the future use of AI must be guided by humanistic and ethical principles. This edited book contributes to the knowledge of how AI can be responsibly and effectively incorporated into educational and research practices that foster innovation, inclusivity, and lifelong learning.

Ho Chi Minh City, Vietnam — Hung Phu Bui
Hue, Vietnam — Le Pham Hoai Huong

# Contents

# Editors and Contributors

## About the Editors

**Hung Phu Bui** is an associate professor at Faculty of Foreign Languages, Ton Duc Thang University, Vietnam. He holds a Ph.D. in English Language Education. His research interests span across many aspects of applied linguistics and language education: second language acquisition, L2 classroom assessment, teaching English for specific purposes, and computer-assisted language teaching and learning. Influenced by educational, linguistic, and psychological perspectives, his endeavors, mainly published in leading journals in the fields of language education and applied linguistics, have been stimulating discussions. Serving as the keynote and plenary speaker in many national and international conferences, Hung has had opportunities to spread his knowledge and research interests to students, colleagues, and novice researchers.

**Le Pham Hoai Huong** is an associate professor in the English Department at Hue University of Foreign Languages and International Studies, Hue University, Vietnam. Her research interests include sociocultural theory, tool-mediated language learning, and language education curriculum development.

## Contributors

**Hariharasudan Anandhan** Department of Language, Culture and Society, College of Engineering and Technology, SRM Institute of Science and Technology, Kattankulathur, Tamil Nadu, India

**Hung Phu Bui** Faculty of Foreign Languages, Ton Duc Thang University, Ho Chi Minh, Vietnam

**Trang Thi Thu Bui** University of Economics Ho Chi Minh City, Ho Chi Minh City, Vietnam;
Tra Vinh University, Vinh Long, Vietnam

**Hoa Thi Hoang Chau** Tra Vinh University, Vinh Long, Vietnam

**Thuy Thanh Dao** Faculty of Social Science, Hong Duc University, Thanh Hoa, Vietnam

**Hoang Thanh Dung** Industrial University of Ho Chi Minh City, Ho Chi Minh City, Vietnam

**M. Gopinath** Sri Krishna Arts and Science College, Coimbatore, Tamil Nadu, India

**Nguyen Thi Bich Hanh** Ha Tinh University, Hà Tĩnh, Vietnam

**Ho Thi Hien** Hue University of Foreign Languages and International Studies, Hue, Vietnam;
Industrial University of Ho Chi Minh City, Ho Chi Minh City, Vietnam

**Le Pham Hoai Huong** Hue University of Foreign Languages and International Studies, Hue University, Hue, Vietnam

**Bünyami Kayali** Technical Sciences Vocational School, Bayburt University, Bayburt, Türkiye

**Tan Hoai Kim** School of Southern Khmer Language, Culture, Arts and Humanities, Tra Vinh University, Vinh Long Province, Vietnam

**Thanh Tien Nguyen** Hoa Sen University, Ho Chi Minh, Vietnam

**Mari Carmen Gil Ortega** Department of Education, University of Bath, Bath, England

**Nguyen Truong Sa** Industrial University of Ho Chi Minh City, Ho Chi Minh City, Vietnam

**Huynh Truong Sang** School of Foreign Languages, Can Tho University, Can Tho, Vietnam;
Faculty of Foreign Languages, An Giang University, Long Xuyen, Vietnam;
Vietnam National University Ho Chi Minh City, Ho Chi Minh, Vietnam

**K. Saritha** Karpagam Academy of Higher Education, Coimbatore, Tamil Nadu, India

**S. Selvalakshmi** Karpagam Academy of Higher Education, Coimbatore, Tamil Nadu, India

**D. Solomon Paul Raj** Sri Krishna Arts and Science College, Coimbatore, Tamil Nadu, India

**Phan Thi Thanh Thao** University of Foreign Languages and International Studies, Hue University, Hue, Vietnam

**Nguyen Thi Diem Thi** Industrial University of Ho Chi Minh City, Ho Chi Minh City, Vietnam

**Nguyen Van Huy** Hue University of Foreign Languages and International Studies, Hue University, Hue, Vietnam

**Nguyen Van Loi** School of Foreign Languages, Can Tho University, Can Tho, Vietnam

**Mehmet Yavuz** Distance Education Application and Research Center, Bingöl University, Bingöl, Türkiye

# Conceptual and Policy Foundations of AI in Education and Research

# Generative Artificial Intelligence (GenAI) in Language Education: Current Trends and Perspectives

**Trang Thi Thu Bui and Hoa Thi Hoang Chau**

**Abstract** Generative artificial intelligence (GenAI) has been gaining a proliferting interest in language education. Education paradigms, influenced by GenAI, have been changing rapidly. Recent research has investigated three main trends that shape language classroom practices. First, AI tools enable the provision of immediate detailed feedback on students' performance. In other words, learners receive instant, objective comments which support self-assessment and language modification on an autonomous basis. Second, GenAI creates new opportunities for sustained exposure to the target language. Learners can engage in open ended dialogues, rehearse tasks, explore content at different levels, and access a wider range of input that matches individual learner variables. This personalized learning modality shows the potential to improve interaction competence and build a more flexible command of language. The third trend concerns AI-assisted assessment. Although a few experts have raised concerns about the integrity of AI-assissted assessment, this innovative approach indicates beneficial for teachers and learners in terms of practicality and rater reliability. Despite promising results from recent studies, remaining concerns about validity, fairness, and transparency require further investigations. This chapter aims to contributes to provide additional resources to the existing literature by critically reviewing the current trends and concepts in AI-assisted language teaching and learning. Although AI-assisted language teaching and learning has bridged the gaps between language education and technology. The remaining concerns may require interdisciplinary research. The ongoing evolution of AI will hopefully develop further, suggesting additional innovations in language education.

**Keywords** AI-assisted language learning · GenAI · Feedback · Exposure to target language · Language assessment

T. T. T. Bui
University of Economics Ho Chi Minh City, Ho Chi Minh City, Vietnam
e-mail: buithithutrang2002@ueh.edu.vn; btttrang2504@sdh.tv.edu.vn

T. T. T. Bui · H. T. H. Chau (✉)
Tra Vinh University, Vinh Long, Vietnam
e-mail: cthhoa@tvu.edu.vn

 
H. P. Bui and L. P. H. Huong (eds.), *Artificial Intelligence in Education and Academic Research*, Studies in Computational Intelligence 1269,
https://doi.org/10.1007/978-3-032-18392-7_1

## 1 Generative Artificial Intelligence

Artificial intelligence (AI) is referred to as the computer's ability to imitate human intelligence (Sheikh et al., 2023). AI makes use of data and algorithms to perform tasks, spanning from simple tasks like stating general knowledge to suggesting possible solutions to a set of related problems (Joksimovic, 2023; Vartiainen et al., 2024). Inspired by AI's ability to perform a task, researchers and practitioners have made several attempts to apply it in many disciplines, including education.

Generative AI (GenAI), such as ChatGPT and Gemini, is a subset of AI. It can create content and images like text (both spoken and written), symbols, and music (Moorhouse & Wong, 2025). The extant literature shows that GenAI has greatly revolutionized education, enforcing enthusiasm and curiosity about AI-supported learning and teaching (Shahzad et al., 2025). However, researchers have raised concerns that the applicability of AI-driven education has been overly decorated (Giannakos et al., 2024).

Despite positive findings from recent research on using GenAI in education, an extensive survey of literature shows that it has several shortcomings. The first concern concentrates primarily on the accuracy of generated content (Li et al., 2025a, 2025b). For instance, while Heintz et al. (2022) found that the language produced by Wordvice is native-like; nevertheless, Özçelik (2023) reported from his study a concern about the accuracy of AI-generated content although it powers proofreading, including spelling and grammar. Second, GenAI is based on the training of large data, suggesting the possibility of included biases (Wang et al., 2024; Wei et al., 2025). As such, it may be essential for AI users to be aware of the ethical guidelines, particularly on protecting privacy and confidentiality. Third, some scholars are concerned about the originality of the works. In the educational setting, students are required to declare that their submissions are original. Declaring that an AI-generated text is one's work is an example of dishonesty as it violates the principle of fairness and academic misconduct in assessment (Luo, 2024; Ortiz-Bonnin et al., 2025). It is a convention in education that students should bear accountability for the contents of their own works (Gonsalves, 2024). Finally, overreliance on GenAI may reduce students' creativity, cognition, and problem-solving and decision-making abilities (Zhai et al., 2024). Such behaviors as copying and pasting AI-created contents may challenge higher-order thinking and creativity. What is education for if it does not inculcate these competencies in students?

GenAI has shifted language education rapidly and greatly. It offers personalized learning, learner autonomy, and interaction opportunities (Kohnke et al., 2025; Szabó & Szoke, 2024). Accordingly, GenAI provides real-time feedback, offers opportunities for language practices, and exposure to authentic language (Tafazoli, 2024). As aforesaid, GenAI enables the generation of texts and images, learners can enter prompts to request a sample text of a particular genre and tasks tailored to their proficiency levels to avail their beyond-class learning.

## 2 Main Trends in Using GenAI in Language Education

An extensive survey of literature shows that there are several research interests in GenAI in language education. They are mainly groups into three main research strands: AI feedback on speaking and writing (e.g. Du & Daniel, 2024; Wang, 2024), AI-supported exposure to authentic language and interaction (e.g. Darmawansah et al., 2025; Zhai & Wibowo, 2023), and AI-assisted language assessment (e.g. Chen et al., 2024; Luc Ha & Nguyen, 2025; Trajkovski & Hayes, 2025). These interests aim to provide learners with enormous opportunities for beyond-class autonomous learning in that AI affordances facilitate, enhance, and transform language education.

One well-documented interest centers on the use of AI feedback, having revolutionized how students refine their language from human–computer interaction (Du & Daniel, 2024). Unlike teacher feedback which requires time and is subjective in nature, AI feedback is instant and objective (Alshehri, 2025; Venter et al., 2025). It is the consistency and immediacy of feedback that can assist students in modifying their language in real time (Yuan, 2025). The feedback provided by GenAI can span from mechanical errors such as typos (Guo & Wang, 2024; Mahapatra, 2024) to language and composition style for which assessment is mainly based on appropriacy (Marzuki et al., 2023; Shi & Aryadoust, 2024). AI systems may track students' progress by comparing students' works and achievements recorded in a particular system, providing students with tailored guidance that meets specific learning needs (Khan et al., 2021; Wang, 2025a, 2025b).

Feedback on oral performance would be somewhat complex as it requires speech recognition, pronunciation modelling, interaction patterns (Chen, 2011; Zou et al., 2023). In second language education, students' mispronunciation can result in misrecognition or no recognition of sounds (Al-khresheh, 2024). Therefore, AI feedback on second language students' speaking can be inaccurate (Shufang, 2021). This may explain why there is less empirical research on AI feedback in L2 speaking than in writing. Hopefully, the ongoing evolution of GenAI suggests the future of AI that feedback may signal or respond to L2 mispronunciation as no pronunciation inaccuracy or unintelligibility.

AI feedback in L2 writing is acknowledged as a proliferating interest compared to the existing body of AI feedback research. Relying on automated writing evaluation, AI has been trained on large data to evaluate on specified criteria (Ding et al., 2025; Khalifa & Albadawy, 2024). According to Aldosemani et al. (2023), the mechanisms in automated writing evaluation generally encompasses feedback which is provided through a computerized system based on large data training (Gao, 2021). AI-powered writing evaluation and text scoring exhibit the possibility of their varying effectiveness levels of grading and feedback (Zhang, 2021). Although it is critiqued for restricted nuanced feedback, automated writing evaluation underscores convenience, immediacy and revision. Tang and Wu (2017) proposed, from a systematic critique of the nature of automated writing evaluation, that pedagogical interventions cannot be missing in terms of writing areas to improve. AI tools, therefore, "can supplement the writing activity without replacing the role of instructor feedback" (Aldosemani et al., 2023).

To request feedback, students have to use prompts specified on particular areas (e.g. coherence and cohesion) of their papers in the hope that they can use feedback to revise their works before official submission to their teacher (Evmenova et al., 2024; Yeung, 2025). It is noteworthy to mention that different AI tools may be powerful in different aspects (Heidt, 2025). As such, students need to use appropriate tools to receive detailed, comprehensive, and reliable feedback (Escalante et al., 2023; Garrel & Mayer, 2024).

However, critics argue that AI feedback may not make sense of contextual issues necessary for nuanced comments (Kotlyar & Krasman, 2025; Tensen et al., 2025). Therefore, AI feedback represents a powerful step forward in language education, but it would be more powerful when incorporated into human judgment and pedagogical expertise (Henderson et al., 2025; Ruwe & Mayweg-Paus, 2023). Unlike feedback given by teachers who can take into account individual learner differences and therefore are able to scaffold students' uptake, comments provided by AI are generally objective regardless of learner variables (Bouziane & Bouziane, 2024; Bui and Barrot, 2024). Considering the accuracy of AI feedback, some scholars (e.g. Guo et al., 2024; Naddaf, 2025) argue that AI tools can be considered peers. As such, their feedback can be used as a facilitative tool for learning rather than an evaluative tool which students can heavily depend on.

A critical central question may be whether AI feedback can enhance students' language competence. Most, if not all, scholars agree that AI can facilitate second language acquisition (Dou et al., 2025; Li et al., 2025a, 2025b; Xiang, 2024). We would argue that the effectiveness of AI feedback regarding learners' language development relies mainly on three issues: what AI tool(s) used by students, how students use the AI tool(s), and what students learn from AI feedback. It is worth noting that different AI tools may provide different types of feedback (Shi & Aryadoust, 2024; Yin et al., 2025). It is the integrity of feedback that could assist them in revising their paper, and comprehensive feedback may support students' revision greatly (Escalante et al., 2023; Lingard, 2023; Olsen & Hunnes, 2024; Su et al., 2023). That is why students need to care about the comprehensiveness, reliability, and validity of feedback given by the AI tool which they intend to use. Also, how students use the AI tool(s) accessible to them may satisfy their expectations. It is advisable that the prompts used by students be as detailed and informative as possible. The more details they provide in their prompts, the more detailed the feedback may be. Their specifications of their feedback are likely to frame AI feedback (Knoth et al., 2024; Lee & Palmer, 2025; Sawalha, 2024). If they use a particular AI tool for a continued manner, they should not forget to prompt AI repeatedly to prevent the effect of AI fatigue on the quality of feedback (Hyland, 2025). Finally, when AI feedback is accessible, it is the students who decide on their language development. Whether and how they integrate the feedback into their finished works to submit to their teachers may give them credit (Henderson et al., 2019; Hyland, 2019; Young & Carless, 2024). However, if they simply revise the "surface" of their works and submit, the chance for language development may be confined. It is what students learn from the feedback. Using feedback to revise a particular paper orients task completion, but learning from feedback to avoid making the same errors in the future and modifying language can crave the path for one's language to develop further (Jackson & Park, 2020; Zhang & Zhang, 2023). AI feedback, in other words, is assistive, but students are decisive in their language development.

The second interest found in the extant literature is the use of GenAI to increase learners' exposure to authentic language. Motivated by the ability of GenAI to generate texts, a few researchers (e.g. Dai & Wu, 2025; Ericsson & Johansson, 2023; Lai & Lee, 2024) have employed AI-assisted systems to provide learners with out-of-class exposure to conversational simulations, showing the importance of this AI use in L2 contexts where exposure to the target language is restricted. GenAI modality tools offer rich input and output illustrated by visuals so that learners can form associations between vocabulary and images or speech and a text or a video clip (Mohammadi et al., 2025, Smith et al., 2025), which proves beneficial for language memory and retention (Hung, 2017, 2019; Mayer, 2014; Paas & van Merriënboer, 2020). Recent experiments have provided empirical findings that learners' prompt literacy and the multimodality of AI have scaffolded students' uptake of the target language and facilitate their interactions (Kim et al., 2025; Tour & Zadorozhnyy, 2025).

The third interest focuses mainly on the use of GenAI in language assessment. Aiming to reduce the workload for teachers, a few researchers have proposed several approaches to developing AI-assisted assessment for classrooms. For instance, Shen et al. (2025) proposed using GenAI to create reading texts for the language class. In the same vein, Wen and Chu (2025) suggested using GenAI to create reading questions. It is important to bear in mind that teachers need to predetermine what they want to test, how to test it (Brown & Abeywick, 2019) and what AI tool they should use (Shen et al., 2025). Teachers may need to consider the content of the text, learners' proficiency level, and test format as they are closely associated with five principles of language assessment: practicality, reliability, validity, authenticity, and washback (Brown & Abeywick, 2019). After considering what to test and how to test it, it may be essential that teachers consider what AI tool(s) should be used to develop an expected test with predetermined specifications. As GenAI can generate texts from specified parameters, prompts should be detailed (e.g. proficiency level and discipline). As different AI tool(s) may have different purposes and yield different results, using the right AI tool(s) can generate a text as the teachers have expected (Shen et al., 2025; Wen & Chu, 2025).

## 3 Conclusion

The ongoing evolution of GenAI has set a pathway for the transformation of language education. The paradigm of language education has been shifted and powered ever by computerization, bridging the gaps between technology and education. The contemporary literature shows that AI has pwoered language learning and teaching enormously, providing learners with autonomy, accessible interaction opportunities, and beyond-class exposure to the target language. Accordingly, learners can enjoy adaptive, personalized, and accessible education, faciliating their engagement in learning and human–computer interaction. Immediate feedback is provided, tailored to learner variables. As AI enables creating texts and converting texts to speech and vice versus, it has drawing researchers' attention to three main trends: AI feedback, AI-powered practice and exposure to the target language, and AI-assisted language assessment. From critical reviews of the mechanisms of AI and potential of AI-powered learning, most, if not all, experts, have proposed that AI can assist learning and teaching, but it cannot replace teachers. However, the rapid development of AI tools have been gearing up the revolution of education ever, so who knows the future of AI-assisted language learning?

**Acknowledgements** This publication was funded by University of Economics Ho Chi Minh City, Vietnam.

**Conflict of Interest** The authors have no conflict regarding this publication.

**Ethics** This manuscript is original. It has not been published anywhere or is under consideration for publication.

## References

Aldosemani, T., Assalahi, H., Lhothali, A., & Albsisi, M. (2023). Automated writing evaluation in EFL contexts. *International Journal of Computer-Assisted Language Learning and Teaching, 13*(1). https://doi.org/10.4018/IJCALLT.329962.

Al-khresheh, M. H. (2024). Phonetic challenges in English: the impact of mispronunciation of the bilabial plosive/p/on communication among Saudi EFL learners. *Cogent Arts & Humanities, 11*(1). https://doi.org/10.1080/23311983.2024.2390777.

Alshehri, A. (2025). AI's effectiveness in language testing and feedback provision. *Social Sciences & Humanities Open, 12*, Article 101892. https://doi.org/10.1016/j.ssaho.2025.101892

Bouziane, K. & Bouziane, A. (2024). AI versus human effectiveness in essay evaluation. *Discover Education, 201*(3). https://doi.org/10.1007/s44217-024-00320-6.

Brown, H. D. & Abeywick, P. (2019). *Language assessment: Principles and classroom practices.* Pearson.

Bui, N. M., & Barrot, J. S. (2024). ChatGPT as an automated essay scoring tool in the writing classrooms: How it compares with human scoring. *Education and Information Technologies, 30*, 2041–2058. https://doi.org/10.1007/s10639-024-12891-w

Chen, H. H. (2011). Developing and evaluating an oral skills training website supported by automatic speech recognition technology. *ReCALL, 23*(1), 59–78. https://doi.org/10.1017/S0958344010000285

Chen, A., Zhang, Y., Jia, J., Liang, M., Cha, Y., & Ping Lim, C. P. (2024). A systematic review and meta-analysis of AI-enabled assessment in language learning: Design, implementation, and effectiveness. *Journal of Computer Assisted Learning, 41*(1), Article e13064. https://doi.org/10.1111/jcal.13064

Dai, L., & Wu, F. (2025). An AI-powered conversational system for college students learning English as a second language. *Education and Information Technology*. https://doi.org/10.1007/s10639-025-13640-3

Darmawansah, D., Hwang, G. J., Lin, C. J., et al. (2025). An artificial intelligence-supported GFCA learning model to enhance L2 students' role-play performance, English speaking and interaction mindset. *Educational Technology Research and Development, 73*, 1451–1479. https://doi.org/10.1007/s11423-025-10453-6

Ding, L., Zou, D., & Kohnke, L. (2025). ChatGPT as an automated writing evaluation tool: How students perceive it and how it affects their writing. *Education and Information Technology*. https://doi.org/10.1007/s10639-025-13775-3

Dou, A., Xu, W., Li, X., Zhang, S., & Zhang, J. (2025). Artificial intelligence in language learning. *International Journal of Distance Education Technologies, 23*(1). https://doi.org/10.4018/IJDET.385045.

Du, J., & Daniel, B. K. (2024). Transforming language education: A systematic review of AI-powered chatbots for English as a foreign language speaking practice. *Computers and Education: Artificial Intelligence, 6*, Article 100230. https://doi.org/10.1016/j.caeai.2024.100230

Ericsson, E., & Johansson, S. (2023). English speaking practice with conversational AI: Lower secondary students' educational experiences over time. *Computers and Education: Artificial Intelligence, 5*, Article 100164. https://doi.org/10.1016/j.caeai.2023.100164

Escalante, J., Pack, A., & Barrett, A. (2023). AI-generated feedback on writing: Insights into efficacy and ENL student preference. *International Journal of Educational Technology in Higher Education, 20*, 57. https://doi.org/10.1186/s41239-023-00425-2

Evmenova, A. S., Regan, K., Mergen, R., et al. (2024). Improving writing feedback for struggling writers: Generative AI to the rescue? *Technology Trends, 68*, 790–802. https://doi.org/10.1007/s11528-024-00965-y

Gao, J. (2021). Exploring the feedback quality of an automated writing evaluation system Pigai. *International Journal of Emerging Technologies in Learning, 16*(11), 322–330. https://doi.org/10.3991/ijet.v16i11.19657

Garrel, J., & Mayer, J. (2024). Which features of AI-based tools are important for students? A choice-based conjoint analysis. *Computers and Education: Artificial Intelligence, 7*, Article 100311. https://doi.org/10.1016/j.caeai.2024.100311

Giannakos, M., Azevedo, R., Brusilovsky, P., Cukurova, M., Dimitriadis, Y., Hernandez-Leo, D., & Rienties, B. (2024). The promise and challenges of generative AI in education. *Behaviour & Information Technology, 44*(11), 2518–2544. https://doi.org/10.1080/0144929X.2024.2394886

Gonsalves, C. (2024). Addressing student non-compliance in AI use declarations: Implications for academic integrity and assessment in higher education. *Assessment & Evaluation in Higher Education, 50*(4), 592–606. https://doi.org/10.1080/02602938.2024.2415654

Guo, K., & Wang, D. (2024). To resist it or to embrace it? Examining ChatGPT's potential to support teacher feedback in EFL writing. *Education and Information Technologies, 29*(7), 8435–8463. https://doi.org/10.1007/s10639-023-12146-0

Guo, K., Zhang, E. D., Li, D., & Yu, S. (2024). Using AI-supported peer review to enhance feedback literacy: An investigation of students' revision of feedback on peers' essays. *British Journal of Educational Technology, 56*(4), 1612–1639. https://doi.org/10.1111/bjet.13540

Heidt, A. (2025). AI for research: The ultimate guide to choosing the right tool. *Nature, 640*, 555–557. https://doi.org/10.1038/d41586-025-01069-0

Heintz, K., Roh, Y., & Lee, J. (2022). Comparing the accuracy and effectiveness of Wordvice AI Proofreader to two automated editing tools and human editors. *Science Editing, 9*(1), 37–45. https://doi.org/10.6087/kcse.261

Henderson, M., Ajjawi, R., Boud, D., & Molloy, E. (2019). *The impact of feedback in higher education: Improving assessment outcomes for learners.* Springer.

Henderson, M., Bearman, M., Chung, J., Fawns, T., Buckingham Shum, S., Matthews, K. E., & de Mello Heredia, J. (2025). Comparing Generative AI and teacher feedback: student perceptions of usefulness and trustworthiness. *Assessment & Evaluation in Higher Education*, 1–16. https://doi.org/10.1080/02602938.2025.2502582.

Hung, B. P. (2017). Vietnamese students learning the semantics of English prepositions. *GEMA Online® Journal of Language Studies, 17*(4), 146–158. https://doi.org/10.17576/gema-2017-1704-10.

Hung, B. P. (2019). A cognitive linguistic approach to teaching English idioms to EFL students: Experimental results. *3L: The Southeast Asian Journal of English Language Studies, 25(*2), 113–126. https://doi.org/10.17576/3L-2019-2502-09.

Hyland, K. (2019). What messages do students take from teacher feedback? In K. Hyland & F. Hyland (Eds.), *Feedback in second language writing contexts and issues* (pp. 265–284). Cambridge University Press. https://doi.org/10.1017/9781108635547.016.

Hyland, K. (2025). EFL teachers and feedback fatigue: AI to the rescue? *Language Teaching*, 1–17. https://doi.org/10.1017/S0261444825101018.

Jackson, D. O., & Park, S. (2020). Self-regulation and personality among L2 writers: Integrating trait, state, and learner perspectives. *Journal of Second Language Writing, 49*, Article 100731. https://doi.org/10.1016/j.jslw.2020.100731

Joksimovic, S., Ifenthaler, D., Marrone, R., De Laat, M., & Siemens, G. (2023). Opportunities of artificial intelligence for supporting complex problem-solving: Findings from a scoping review. *Computers and Education: Artificial Intelligence, 4*, Article 100138. https://doi.org/10.1016/j.caeai.2023.100138

Khalifa, M., & Albadawy, M. (2024). Using artificial intelligence in academic writing and research: An essential productivity tool. *Computer Methods and Programs in Biomedicine Update, 5*, Article 100145. https://doi.org/10.1016/j.cmpbup.2024.100145

Khan, I., Ahmad, A. R., Jabeur, N., et al. (2021). An artificial intelligence approach to monitor student performance and devise preventive measures. *Smart Learning Environment, 8*, 17. https://doi.org/10.1186/s40561-021-00161-y

Kim, J., Lee, S. S., Detrick, R., et al. (2025). Students-Generative AI interaction patterns and its impact on academic writing. *Journal of Computing in Higher Education.* https://doi.org/10.1007/s12528-025-09444-6

Knoth, N., Tolzin, A., Janson, A., & Leimeister, J. M. (2024). AI literacy and its implications for prompt engineering strategies. *Computers and Education: Artificial Intelligence, 6*, Article 100225. https://doi.org/10.1016/j.caeai.2024.100225

Kohnke, L., Zou, D., & Su, F. (2025). Exploring the potential of GenAI for personalised English teaching: Learners' experiences and perceptions. *Computers and Education: Artificial Intelligence, 8*, Article 100371. https://doi.org/10.1016/j.caeai.2025.100371

Kotlyar, I., & Krasman, J. (2025). Student reactions to AI versus human feedback in teamwork skills assessment. *International Journal of Educational Technology in Higher Education, 22*, 57. https://doi.org/10.1186/s41239-025-00555-9

Lai, W. Y. W., & Lee, J. S. (2024). A systematic review of conversational AI tools in ELT: Publication trends, tools, research methods, learning outcomes, and antecedents. *Computers and Education: Artificial Intelligence, 7*, Article 100291. https://doi.org/10.1016/j.caeai.2024.100291

Lee, D., & Palmer, E. (2025). Prompt engineering in higher education: A systematic review to help inform curricula. *International Journal of Educational Technology in Higher Education, 22*, 7. https://doi.org/10.1186/s41239-025-00503-7

Li, F., Yang, Y., & Yu, G. (2025a). Nudging perceived credibility: The impact of AIGC labeling on user distinction of AI-generated content. *Emerging Media, 3*(2), 275–304. https://doi.org/10.1177/27523543251317572

Li, M., Wang, Y., & Yang, X. (2025b). Can generative AI chatbots promote second language acquisition? A meta-analysis. *Journal of Computer Assisted Learning, 41*(4), Article e70060. https://doi.org/10.1111/jcal.70060

L. Lingard (2023). Writing with ChatGPT: an illustration of its capacity, limitations & implications for academic writers. *Perspectives in Medicine Education, 12*(1), 261–270. https://doi.org/10.5334/pme.1072.

Luc Ha, D. N., & Nguyen, A. T. (2025). Artificial intelligence-based assessment in ELT exam creation: A case study of Van Lang University lecturers. *Saudi Journal of Language Studies, 5*(1), 34–49. https://doi.org/10.1108/SJLS-06-2024-0030

Luo, J. (2024). A critical review of GenAI policies in higher education assessment: A call to reconsider the "originality" of students' work. *Assessment & Evaluation in Higher Education, 49*(5), 651–664. https://doi.org/10.1080/02602938.2024.2309963

Mahapatra, S. (2024). Impact of ChatGPT on ESL students' academic writing skills: A mixed methods intervention study. *Smart Learning Environments, 11*(9), 1–18. https://doi.org/10.1186/s40561-024-00295-9

Marzuki, Widiati, U., Rusdin, D., Darwin, & Indrawati, I. (2023). The impact of AI writing tools on the content and organization of students' writing: EFL teachers' perspective. *Cogent Education, 10*(2). https://doi.org/10.1080/2331186X.2023.2236469.

Mayer, R. E. (2014). Cognitive theory of multimedia learning. In R. E. Mayer (Ed.), *The Cambridge Handbook of Multimedia Learning* (pp. 43–71). Cambridge University Press. https://doi.org/10.1017/CBO9781139547369.005.

Mohammadi, M., Tajik, E., Martinez-Maldonado, R., Sadiq, S., Tomaszewski, W., Khosravi, H. (2025). Artificial intelligence in multimodal learning analytics: A systematic literature review. *Computers and Education: Artificial Intelligence, 8*, 100426. https://doi.org/10.1016/j.caeai.2025.100426.

Moorhouse, B. L., & Wong, K. M. (2025). Generative artificial intelligence and language teaching. *Cambridge University Press*. https://doi.org/10.1017/9781009618823

Naddaf, M. (2025). AI is transforming peer review—and many scientists are worried. *Nature, 639*, 852–854. https://doi.org/10.1038/d41586-025-00894-7

Olsen, T., & Hunnes, J. (2024). Improving students' learning—the role of formative feedback: Experiences from a crash course for business students in academic writing. *Assessment & Evaluation in Higher Education, 49*(2), 129–141. https://doi.org/10.1080/02602938.2023.2187744

Ortiz-Bonnin, S., & Blahopoulou, J. (2025). Chat or cheat? Academic dishonesty, risk perceptions, and ChatGPT usage in higher education students. *Social Psychology in Education, 28*, 113. https://doi.org/10.1007/s11218-025-10080-2

Özçelik, N. P. (2023). A comparative analysis of proofreading capabilities: Language experts vs. ChatGPT. *In International studies in educational sciences*. Serüven Publishing 147–161. https://www.researchgate.net/publication/372237321

Paas, F., & van Merriënboer, J. J. G. (2020). Cognitive-load theory: Methods to manage working memory load in the learning of complex tasks. *Current Directions in Psychological Science, 29*(4), 394–398. https://doi.org/10.1177/0963721420922183

Ruwe, T. & Mayweg-Paus, E. (2023). "Your argumentation is good", says the AI versus humans—The role of feedback providers and personalised language for feedback effectiveness. *Computers and Education: Artificial Intelligence, 5*, 100189. https://doi.org/10.1016/j.caeai.2023.100189.

Sawalha, G., Taj, I., & Shoufan, A. (2024). Analyzing student prompts and their effect on ChatGPT's performance. *Cogent Education, 11*(1). https://doi.org/10.1080/2331186X.2024.2397200.

Shahzad, M. F., Xu, S., An, X., & Asif, M. (2025). Are generative AI technologies transforming education for the 21st Century? Research trends, challenges, and benefits. *Sage Open, 15*(3). https://doi.org/10.1177/21582440251368594.

Sheikh, H., Prins, C., Schrijvers, E. (2023). Artificial intelligence: Definition and background. In H. Sheikh, C. Prins, E. Schrijvers (Eds.), *Mission AI: Research for policy*. Springer. https://doi.org/10.1007/978-3-031-21448-6_2.

Shen, L., Kane-Cabello, J., Candelaria, P. Y., Stratford, D., & Clemens, N. H. (2025). Can artificial intelligence tools to generate text that is useful for reading practice? *Learning Disabilities Research & Practice, 40*(4), 191–204. https://doi.org/10.1177/09388982251352564

Shi, H., & Aryadoust, V. (2024). A systematic review of AI-based automated written feedback research. *ReCALL*, 1–23. https://doi.org/10.1017/S0958344023000265.

Shufang, Z. (2021). Design of an automatic English pronunciation error correction system based on radio magnetic pronunciation recording devices. *Journal of Sensor*. https://doi.org/10.1155/2021/5946228

Smith, B. E., Shimizu, A. Y., Burriss, S. K., Hundley, M., & Pendergrass, E. (2025). Multimodal composing with generative AI: Examining preservice teachers' processes and perspectives. *Computers and Composition, 75*, Article 102896. https://doi.org/10.1016/j.compcom.2024.102896

Su, Y., Lin, Y., & Lai, C. (2023) Collaborating with ChatGPT in argumentative writing classrooms. *Assessment in Writing, 57* (2023), 100752. https://doi.org/10.1016/j.asw.2023.100752.

Szabó, F., & Szoke, J. (2024). How does generative AI promote autonomy and inclusivity in language teaching? *ELT Journal, 78*(4), 478–488. https://doi.org/10.1093/elt/ccae052

Tafazoli, D. (2024). Exploring the potential of generative AI in democratizing English language education. *Computers and Education: Artificial Intelligence, 7*, Article 100275. https://doi.org/10.1016/j.caeai.2024.100275

Tang, J., & Wu, Y. (2017). Impact of AWE rubrics and automated assessment on EFL writing instruction. *International Journal of Computer-Assisted Language Learning and Teaching, 7*(2), 58-s74. https://doi.org/10.4018/IJCALLT.2017040104

Tensen, D., Grainger, P., & Graham, W. (2025). Using AI to generate formative feedback in doctoral education. *Assessment & Evaluation in Higher Education*, 1–17. https://doi.org/10.1080/02602938.2025.2536558.

Tour, E., & Zadorozhnyy, A. (2025). Conceptualizing and operationalizing prompt literacy for English language learners. *Journal of Adolescent & Adult Literacy, 69*(3), Article e70020. https://doi.org/10.1002/jaal.70020

Trajkovski, G. & Hayes, H. (2025). *AI-assisted assessment in education: Transforming assessment and measuring learning*. Springer.

Vartiainen, H., Kahila, J., Tedre, M., López-Pernas, S., & Pope, N. (2024). Enhancing children's understanding of algorithmic biases in and with text-to-image generative AI. *New Media & Society, 27*(9), 5342–5368. https://doi.org/10.1177/14614448241252820

Venter, J., Coetzee, S. A., & Schmulian, A. (2025). Exploring the use of artificial intelligence (AI) in the delivery of effective feedback. *Assessment & Evaluation in Higher Education, 50*(4), 516–536. https://doi.org/10.1080/02602938.2024.2415649

Wang, Y. (2025a). Artificial intelligence in student management systems to enhance academic performance monitoring and intervention. *Scientific Reports, 15*, 35122. https://doi.org/10.1038/s41598-025-19159-4

Wang, C. (2025b). Exploring students' generative AI-assisted writing processes: Perceptions and experiences from native and nonnative English speakers. *Technology, Knowledge, and Learning, 30*, 1825–1846. https://doi.org/10.1007/s10758-024-09744-3

Wang, H., Dang, A., Wu, Z., & Mac, S. (2024). Generative AI in higher education: Seeing ChatGPT through universities' policies, resources, and guidelines. *Computers and Education: Artificial Intelligence, 7*, Article 100326. https://doi.org/10.1016/j.caeai.2024.100326

Wei, X., Kumar, N., & Zhang, H. (2025). Addressing bias in generative AI: Challenges and research opportunities in information management. *Information & Management, 62*(2), Article 104103. https://doi.org/10.1016/j.im.2025.104103

Wen, Z., & Chu, S. K. W. (2025). Using generative AI for reading question creation based on PIRLS 2011 framework. *Cogent Education, 12*(1), 2458653. https://doi.org/10.1080/2331186X.2025.2458653

Xiang, C. (2024). How AI helps second language learners acquire target language: An analysis from a cross-cultural perspective. *Transactions on Social Science, Education and Humanities Research, 11*, 406–411. https://doi.org/10.62051/mjfh7285

Yeung, S. (2025). University students' engagement with generative AI-supported automated writing evaluation (AWE) feedback. *Journal of Second Language Writing, 68*, Article 101203. https://doi.org/10.1016/j.jslw.2025.101203

Yin, J., Xu, H., Pan, Y., & Hu, Y. (2025). Effects of different AI-driven Chatbot feedback on learning outcomes and brain activity. *NPJ Science of Learning, 10*, 17. https://doi.org/10.1038/s41539-025-00311-8

Young, S., & Carless, D. (2024). Investigating variation in undergraduate students' feedback seeking experiences: Towards the integration of feedback seeking within the curriculum. *Assessment & Evaluation in Higher Education, 49*(8), 1048–1060. https://doi.org/10.1080/02602938.2024.2338537

Yuan, H. (2025). Artificial intelligence in language learning: biometric feedback and adaptive reading for improved comprehension and reduced anxiety. *Humanities & Social Sciences Communications, 12*, 556. https://doi.org/10.1057/s41599-025-04878-w.

Zhai, C., & Wibowo, S. (2023). A systematic review on artificial intelligence dialogue systems for enhancing English as foreign language students' interactional competence in the university. *Computers and Education: Artificial Intelligence, 4*, Article 100134. https://doi.org/10.1016/j.caeai.2023.100134

Zhai, C., Wibowo, S., & Li, L. D. (2024). The effects of over-reliance on AI dialogue systems on students' cognitive abilities: A systematic review. *Smart Learning Environment, 11*, 28. https://doi.org/10.1186/s40561-024-00316-7

Zhang, S. (2021). Review of automated writing evaluation systems. *Journal of China Computer-Assisted Language Learning, 1*(1), 170–176. https://doi.org/10.1515/jccall-2021-2007

Zhang, X., & Zhang, R. (2023). Feedback, response, and learner development: A sociocultural approach to corrective feedback in second language writing. *Sage Open, 13*(1). https://doi.org/10.1177/21582440231157680.

Zou, B., Du, Y., Wang, Z., Chen, J., & Zhang, W. (2023). An investigation into artificial intelligence speech evaluation programs with automatic feedback for developing EFL learners' speaking skills. *Sage Open, 13*(3), https://doi.org/10.1177/21582440231193818.

**Bui Thi Thu Trang** is a lecturer at School of Foreign Languages, University of Economics Ho Chi Minh City (UEH), Vietnam. She holds an MA degree in Applied Linguistics from La Trobe University, Australia. She is currently a PhD candidate in TESOL at Tra Vinh University (TVU), with research interests in AI, scaffolding, feedback, student motivation, learner autonomy and ELT methodology.

**Hoa Thi Hoang Chau** is the Director of International Affairs, Tra Vinh University, Vietnam. She earned her Ph.D. in TESOL from Hue University and her M.Ed. and B.Ed. in TESOL from Can Tho University. Her research focuses on English language education and teacher training and EMI and online learning and cultural integration in EFL classrooms to develop intercultural communicative competence.

# AI Integration into Higher Education Curricula: A Multiple-Case Study of Key Universities in Vietnam

**Le Pham Hoai Huong**

**Abstract** With the aim of exploring how universities in Vietnam incorporate artificial intelligence (AI) into higher education curricula, this qualitative study is undertaken to examine the extent to which eight leading institutions respond to Vietnam's strategy for AI integration in education and research by 2030 (https://en.baochinhphu.vn/national-strategy-on-rd-and-application-of-artificial-intelligence-11140663.htm). This multiple-case study collects data from multiple sources, including institutions' documents, government policies, research publications, and media reports for eight universities in Vietnam. The thematic analysis indicates that AI is embedded in interdisciplinary subjects and ethical considerations for AI use are being formulated. It is also found that key universities, such as Vietnam National University-Hanoi and RMIT Vietnam align well with the country's strategy on AI use by 2030 and other regional institutions integrate AI in their curricula within resource constraints. The findings shed light on policy-to-practice processes of integrating AI into curricula and emphasize the importance of following contextualized, ethically grounded, and interdisciplinary approaches for AI integration in higher education programs.

**Keywords** Artificial intelligence · Curriculum innovation · Vietnam

## 1 Introduction

In 2021, the government of Vietnam issued the national strategy on research, development and application of artificial intelligence until 2030. It is a framework for incorporating AI in research and application, promoting human capital, and modernizing Vietnamese higher education (Government of Vietnam, 2021). This necessitates systemic changes within the institutions in Vietnam. Thus, exploring the extent

L. P. H. Huong (✉)
Hue University of Foreign Languages and International Studies, Hue University, Hue, Vietnam
e-mail: lphhuong@hueuni.edu.vn

 
H. P. Bui and L. P. H. Huong (eds.), *Artificial Intelligence in Education and Academic Research*, Studies in Computational Intelligence 1269,
https://doi.org/10.1007/978-3-032-18392-7_2

to which AI incorporation transforms higher education practices, feedback mechanisms, and data management (Crompton & Burke, 2023) is of practical significance. By focusing on the key universities in Vietnam and their initiatives in adopting AI in their higher education curricula, the study's findings can provide practical implications for the country's policy framework related to AI integration.

The study addresses two research questions: (1) How do universities in Vietnam integrate AI into their higher education curricula? and (2) To what extent does this integration align with Vietnam's AI strategy by 2030?

# 2 Literature Review

## *2.1 AI in Education and Research*

### 2.1.1 Affordances

In education, AI tools are being adopted to support instruction, assessment, and research workflows, thereby enabling educational innovations. AI is also being used to "customize prompt feedback, adapt instruction, and predict students' learning success" (Crompton & Burke, 2023, p. 2). Furthermore, the emergence of AI has shifted educators' roles in the classroom, enhancing their capabilities, especially in administrative and routine pedagogical tasks. "AI can free up teachers from repetitive grading and content delivery" (Chan & Tsi, 2023, p. 5). This transformation requires educators to work more as facilitators rather than transmitters of knowledge in faculty training and institutional policy. Other affordances of AI lie in academic research. It is considered to provide potential for enhancing research, for example, literature review and data analysis (Alqahtani et al., 2023).

### 2.1.2 Constraints

Despite the benefits, AI use raises questions about academic integrity and assessment design. The use of AI in education and research raises ethical, practical, and epistemological issues. Embedding AI into curricula must consider pedagogical intentionality and clear guidelines to avoid misuse or overdependence (Williams, 2025). Thus, its use requires careful governance, ethical considerations, and critical engagement by all stakeholders in the academic community.

## 2.2 *AI Policy in Vietnam*

### 2.2.1 The Nation's Strategy for AI Integration

In 2021, the government of Vietnam issued the AI strategy by 2030 which provides guidelines for the country to integrate AI across multiple disciplines ranging from education, research, and socio-economic development. The aim is to enhance the capacity of production in these fields and the country's ability to compete with other ASEAN countries sustainably (Government of Vietnam, 2021, p. 1). A central focus of the strategy is to develop AI-related education and research infrastructure. It is stated in the official document that by 2025, the government will have established "two national innovation centers for AI" and "upgrade and form 10 new key research and training institutions on AI" (Government of Vietnam, 2021, p. 2). The long-term goal is to position Vietnam among the leading countries in ASEAN in AI research, development and application by the year 2030 (Government of Vietnam, 2021, p. 2).

Furthermore, the strategy states that the country emphasizes building human capacity by using basic skills in AI application and data science for young people's creativity as well as offering training courses for short and medium terms for workers for their career transitions (Government of Vietnam, 2021, p. 3). In addition, the government has planned to build key centers for research and training personnel in AI in leading universities and research institutions (Government of Vietnam, 2021, p. 3). These educational efforts also target technical training and fostering a creative society *and* effective government that uses AI to enhance public administration and service delivery (Government of Vietnam, 2021, p. 2).

### 2.2.2 Vietnamese Ministry of Education and Training (MOET)'s Role in AI Implementation

The Vietnamese MOET is implementing a key role in realizing mass programs on popularizing data building skills, applying AI for the youth, promoting formal training programs on AI, data science, and training subjects on data analysis, AI application in the curriculum of different disciplines in universities and colleges (Government of Vietnam, 2021, p. 6). Furthermore, it is directed to promote AI applications in education, envisioning the job prospects, automating teacher professional development, and improving learning efficiency (Government of Vietnam, 2021, p. 6). This national strategy provides a comprehensive policy framework for Vietnamese universities to align their curricula with the government's goals.

### 2.2.3 Research on AI Use in the Vietnamese Context

Recent studies have explored how AI is being integrated into educational institutions in Vietnam. For example, Nguyen (2023) found that AI applications in Vietnam can be classified into three domains: guidance, teacher, and student. In the first category, AI-enabled guidance system in Vietnam are to match students with suitable teachers and provide early warnings for at-risk students. In the next domain, AI tools assist teachers thanks to such applications as automated essay scoring and real-time dashboards support educators in grading and identifying students who may require additional support. Regarding the last domain, AI applications often include adaptive learning platforms and game-based learning environments to promote individualized learning experiences and enhance engagement.

Previous studies have delved into AI use from a policy and institutional perspective. For instance, Quy et al. (2023) present a case study of Hung Yen University of Technology and Education case study, showcasing how it is conducting digital transformation. The authors point out that AI is used not only for learning personalization but also for administrative forecasting in such tasks as predicting student enrollment, teaching needs, and resource allocation. However, the authors acknowledge the challenges related to infrastructural and financial issues. Generally, AI has been found to be integrated into the education system and the outcomes seem to enhance learning, and support institutional management. Nonetheless, there exist challenges such as ethical concerns related to data use, and limited human resources.

## 2.3 *Educational Change Framework by Fullan (2015)*

This study adopts the educational change framework (Fullan, 2015) because the embedding of AI into higher education programs requires adaptaption. Originally, the framework views sustainable reform as relying on combining theory (meaning) and practice (action) so that stakeholders find meaning from shared efforts, recognizing that change in beliefs and understanding is the foundation for success across the entire system (Fullan, 2015).

The framework has been developed further and indicates a process-oriented lens for understanding systemic reform (Fullan, 2015). In other words, educational transformation should be regarded as an interactive process influenced by professional learning, leadership engagement, and organizational culture. These principles are particularly relevant to policies related to AI embedding in tertiary education, where institutional readiness and teacher competence determine the success of technology adoption.

In the Vietnamese context, where AI initiatives are driven by national directives under the national strategy on AI to 2030, Fullan's model is able to explain the contextual and political dynamics influencing AI adoption in higher education curricula as well as reviewing the shared efforts among the parties to bring the change across the educational system.

# 3 Methodology

## 3.1 Reserach Design

The current study is designed following the principles of a qualitative multiple-case study. It aims to investigate how AI has been embedded in higher education curricula across the key Vietnamese universities. This approach is appropriate because of the diversity of institutional missions, educational models, and disciplinary foci in Vietnam's higher education system. Besides, such a design allows for cross-case comparison. Yin (2018) states that a multiple-case study enables the identification of cross-institutional patterns, and context-specific insights.

## 3.2 Case Sampling

To implement the multiple-case study, purposive sampling is used with eight universities representing a broad context of Vietnam's higher education. The sample includes:

- Can Tho University (CTU)
- Fulbright University Vietnam
- Hue University
- RMIT University Vietnam
- Thai Nguyen University (TNU)
- The University of Danang (UD)
- Vietnam National University, Hanoi (VNU-HN)
- Vietnam National University, Ho Chi Minh City (VNU-HCMC)

The eight institutions vary in educational training programs ranging from public and private to international institutions. Besides, they are located in different regions of northern, central, and southern Vietnam. These institutions offer different disciplinary foci including technical, liberal arts, and interdisciplinary subjects. They have demonstrated their leadership in digital transformation, involvement in national AI strategy, and innovation in curriculum development.

## 3.3 Data Collection and Analysis

### 3.3.1 Data Collection

The study's data sources are derived from document-based triangulation from multiple credible and publicly available secondary sources. The four main types of data include institutional documentation, public communications, national policy frameworks, and academic and media sources. The first source includes university websites, program handbooks, academic catalogs, and strategic plans which are examined to identify how AI is integrated into curricular structures, for example, degree programs, elective modules, interdisciplinary offerings, or inclusion in pedagogical and ethical training. The second source is public communications such as conference proceedings (e.g., VNU AI Day), university press releases, and MOET partnership announcements for understanding of these institutions' strategic goals related to AI adoption. Thirdly, institutions' curricular developments are cross-checked with national initiatives such as Vietnam's national strategy on research, development and application of AI by 2030 and MOET's digital education reform guidelines for evaluation of policy alignment and institutional responsiveness to government directives. Finally, scholarly articles (e.g., from the VNU Journal of Science *or* Vietnam Journal of Education and Technology) and reputable media outlets (e.g., Tuổi Trẻ, Lao Động) are reviewed to explore external evaluations, student perspectives, and broader discourse on the application of AI in higher education curricula. All data sources are then triangulated for validity and understanding of institutional, policy, academic, and societal perspectives.

#### 3.3.2 Data Analysis

This study adopts qualitative content analysis to interpret and categorize diverse approaches to AI adoption used by the institutions used. Adopting Schreier's (2012) analysis principles, the data are examined inductively, searching for patterns, themes, and conceptual dimensions that emerge from the data sources. This approach allows for analytical flexibility while maintaining thematic coherence. Table 1 below outlines the four key analytical dimensions.

## 4 Findings

### *4.1 The Case of Can Tho University*

Can Tho University (CTU) is located in the Mekong Delta region of Vietnam. It is a well-established institution with 121 undergraduate and 58 master's programs. It serves over 57,000 students in multidisciplinary fields and is committed to teaching, scientific research, and technology transfer (Can Tho University, n.d.). As Vietnam advances its national strategy for AI to 2030, CTU is actively positioning itself as a leader in AI education and research in the region.

**Table 1** Analytical dimensions

| Analytical dimension | Description | Example from data sources |
|---|---|---|
| Depth of AI integration | To see whether AI is integrated as a core subject, elective, or peripheral topic within the curriculum | VNU-Hanoi includes compulsory AI modules in first-year undergraduate courses across disciplines (VNU, 2025a, 2025b) |
| Pedagogical scope | To assess the inclusion of AI in terms of technical competencies, ethical reasoning, and interdisciplinary applications | Hue University promotes both AI technical skills and digital ethics through science talks and medical training (University of Sciences—Hue University, 2025) |
| Curricular innovation | To identify the presence of AI-related minors, micro-credentials, and cross-departmental collaborations | Fulbright University Vietnam offers interdisciplinary courses on AI (Fulbright University Vietnam, 2024) |
| Alignment with the nation's strategy | To evaluate institutional alignment with national AI strategies and MOET policy frameworks | The University of Danang's AI training programs reflect MOET's push for regional centers of excellence (The University of Da Nang, 2025a) |

### 4.1.1 AI Integration into Higher Education Curricula

The university's initiatives for AI started in November 2024 when the institution announced the opening of a new major of AI at both the undergraduate and graduate levels (Vietnam.vn, 2024). This formal curricular adoption of AI is part of CTU's broader mission to align with national and global technology trends while preparing a workforce skilled in digital competencies.

Member institutions of CTU also foster AI learning through academic seminars and co-curricular actitivities. For instance, College of Agriculture introduced a seminar titled *The Use of AiI in Scientific Writing* which focused on ChatGPT and Copilot for research activities, editorial tasks, and improving publication quality. It aimed to promote academic writing workflows and support international-level research publications (College of Agriculture—CTU, 2025, para. 3).

Other activities hosted by CTU also support faculty development in AI pedagogy. For example, it held a conference titled *"Higher Education in the Age of AI"* as part of its digital transformation roadmap in June 2025. This event emphasized AI's role in teaching innovation, academic governance, and learning analytics. CTU committed to "invest in infrastructure, improve faculty capacity, and expand international cooperation to ensure effective AI integration in teaching and research" (Can Tho University, 2025, para. 6).

#### 4.1.2 Alignment with Vietnam's AI Strategy and Policy Framework

It can be seen that CTU's AI initiatives align with Vietnam's strategy to 2030 and the MOET's digital transformation policy. To be more specific, the university offers courses on big data as part of a broader shift toward a smart university model (Can Tho University, 2025). In a conference in June 2025, the university leaders discussed institutional regulations for AI use in academia. It aims to balance innovation with academic standards (Can Tho University, 2025, para. 9). As the result of these initiatives, Can Tho city has emerged as one of the centers for AI research, particularly in such areas as machine learning, natural language processing, and AI applications in agriculture and smart cities (Vocal Media, 2024, para. 3).

### *4.2 The Case of Fulbright University Vietnam*

Founded in 2016, Fulbright University Vietnam (FUV) is a liberal arts institution that emphasizes interdisciplinary learning, critical inquiry, and civic engagement (Fulbright University Vietnam, n.d.-a).

#### 4.2.1 AI Integration into Higher Education Curricula

FUV integrates AI into its broader curriculum through both core and elective courses across technical and non-technical disciplines. The university's website indicates that the program courses in engineering and computer science cover fundamental AI concepts in machine learning, data science, and algorithms (Fulbright University Vietnam, 2024). Non-STEM students study AI in public policy and ethics and engage with issues such as algorithmic bias and the governance of digital platforms. It is stated on the website that students have to think critically about AI's influence on decision-making and accountability in government (Fulbright University Vietnam, 2023a, 2023b, para. 2). In addition, the institution implements a cross-disciplinary exploration of AI in applied research addressing real-world challenges. Instructors are required to apply project-based teaching and co-curricular engagement, allowing students to use AI tools in fields ranging from environmental monitoring to civic technology innovation (Fulbright University Vietnam, 2024). Although the university does not currently offer postgraduate AI programs, it is developing its faculty expertise and infrastructure in preparation for future development (Fulbright University Vietnam, n.d.-a).

#### 4.2.2 Alignment with Vietnam's AI Strategy and Policy Framework

The university's website indicates that FUV aligns with the Vietnam's national AI strategy by 2030 and MOET's objectives by embedding AI training within both

STEM and social science education, encouraging students to explore the societal impacts of emerging technologies (Fulbright University Vietnam, 2023a, 2023b). The institution's AI education practices reflect the national emphasis on preparing students not only to become technical experts, but also to be digitally responsible citizens who can contribute to Vietnam's sustainable and inclusive development (Fulbright University Vietnam, 2024).

## *4.3 The Case of Hue University*

Hue University was established in 1957. It has taken initiatives to integrate AI into both its technical and non-technical curricula to build a digitally fluent workforce.

### 4.3.1 AI Integration into Higher Education Curricula

At Hue University, students can enroll in a data science and AI program which was launched in 2020 through its Faculty of Engineering and Technology. This program includes two pathways: a bachelor's track and an engineering track. While the bachelor's curriculum emphasizes research skills and theoretical grounding, the engineering track focuses on applied skills, practical internships, and enterprise collaboration (Hue University, 2020). Students can specialize in one of three concentrations: data science, artificial intelligence, or business data analytics.

In addition, at Hue University, AI has been promoted through co-curricular initiatives such as talk shows and forums. In May 2025, the University of Sciences, a member institution of Hue University hosted an event on emerging trends in AI-supported learning where students explored tools like ChatGPT, Microsoft Copilot, and Gemini. These tools were introduced not just as aids to productivity, but also as platforms to foster critical thinking, scientific reasoning, and academic integrity. The event emphasized that AI does not replace human thinking but serves as a powerful tool helping learners to save time and to promote critical thinking (University of Sciences—Hue University, 2025, para. 7).

In the field of medical education, the University of Medicine and Pharmacy, a key member of Hue University, has partnered with FPT Corporation to implement AI and digital technologies in telemedicine, diagnostics, and medical training. According to the collaboration agreement, "AI will be integrated into remote medical services improving patient experience and medical service quality" (FPT & Hue University of Medicine and Pharmacy, 2024, para. 9). Furthermore, such events as the *Science Café on AI in Medicine* aims to contribute to professional dialogue and student engagement in AI applications for healthcare (HueMed, 2025).

### 4.3.2 Alignment with Vietnam's AI Strategy and Policy Framework

The AI bachelor's program, digital literacy forums, and intersectoral collaborations indicate Hue University's alignment with the national objectives for human capital development, innovation, and interdisciplinary integration. According to Giao Duc Thoi Dai (2025, para. 4), Hue University's faculty leadership has publicly stressed the need to equip students with AI knowledge and skills for study and research. Furthermore, collaboration with enterprises like FPT and the introduction of smart university models show that Hue University is not only aligning with but also operationalizing the AI roadmap laid out in the Vietnam's strategy for AI integration. Overall, the emphasis on building a future for digital ethics for AI users reflects the national calls for responsible technology adoption.

## *4.4 The Case of RMIT Vietnam*

RMIT Vietnam has been considered to be one of the regional leaders in innovative and future-oriented education, particularly in AI-focused education in Vietnam. It is a branch of the Royal Melbourne Institute of Technology (RMIT), founded in 2000. RMIT Vietnam offers internationally recognized programs in business, communication, design, technology, and engineering. Its educational model emphasizes work-integrated learning, applied research, and industry engagement (RMIT Vietnam, n.d.-a).

### 4.4.1 AI Integration into Higher Education Curricula

RMIT Vietnam offers a specialized bachelor of AI program that provides students with knowledge and skills in AI-related fields such as data science, machine learning, and deep learning (RMIT Vietnam, n.d.-b). At RMIT, the AI program is designed for students seeking to develop or advance careers in AI. The course advances computing theory with practical applications and emphasizes ethics and social responsibility. To meet the course requirements, students need to obtain knowledge and skills from the courses, apply, and assess them in a real or simulated workplace setting and receive direct feedback from the industry (RMIT Vietnam, n.d.-c, p. 1). Students may also pursue a research stream culminating in a thesis, which can serve as a pathway to a PhD.

### 4.4.2 Alignment with Vietnam's AI Strategy and Policy Framework

RMIT claims the timely AI incorporation in its educational programs because businesses now use AI in relation to required resources, regulations, ethical matters (RMIT Vietnam, 2020, para. 14). Its growing collaboration with industry also fulfills the national emphasis on linking higher education curricula with socio-economic needs. In general, RMIT Vietnam's AI initiatives respond well to Vietnam's AI policy objectives. Its emphasis on interdisciplinary training, ethics, and applied research resonates with national policy goals that stress human-centered and ethical AI implementation as indicated in its program's design, which incorporates ethical training and real-world simulation (RMIT Vietnam, n.d.-c).

## *4.5 The Case of Thai Nguyen University*

Thai Nguyen University (TNU) was founded in 1994 through the merger of five existing colleges in northern Vietnam. With its mission to be a center for innovation and community development, TNU aims to deliver high-quality education and cutting-edge research for the northern midland and mountainous regions of Vietnam (Thai Nguyen University, n.d.).

### 4.5.1 AI Integration into Higher Education Curricula

Thai Nguyen University has integrated AI into its teaching and training activities. The university has taken significant steps to embed AI tools and develop AI-related competencies into instruction and learning across its member institutions. Two large-scale training sessions in November 2023 and April 2024 equipped over 500 lecturers and academic staff from TNU's member colleges, including the University of Education, the University of Science, and the School of Foreign Languages with skills in AI-based teaching techniques (Thai Nguyen University, 2023, 2024).

The university also organized activities to promote AI adoption and literacy. For example, in a workshop series that the institution held, participants explored such tools as ChatGPT, Canva, and text-to-speech generators, with the goal of improving both instructional productivity and student engagement. In addition, the workshops provided sessions on AI foundations, prompt engineering, and using generative AI for lesson design, data visualization, and content creation (Thai Nguyen University, 2023, para. 6).

Furthermore, TNU facilitated student-focused research on AI integration. A 2024 study at the University of Science (a member institution of TNU) showed that students had a generally positive awareness of AI tools for self-study but expressed concern about ethical use and over-reliance (Bui et al., 2024). The study by Lan and Van (2025) reveals that AI improved writing skills and access to translation tools among students. The students however raised concerns regarding creativity loss, data privacy, and emotional limitations of AI-based platforms.

### 4.5.2 Alignment with Vietnam's Strategy and Policy Framework

TNU's AI seminars and policy implementation efforts reflect MOET's call for responsible and ethical AI use. Although TNU has not yet launched an official AI major, its institutional roadmap supports the progressive adoption of AI across disciplines. These activities are building the foundation for broader, policy-aligned reforms in the near future.

## *4.6 The Case of the University of Da Nang (UD)*

UD is a leading university in central Vietnam, offering innovative education and research aligned with the digital transformation agenda, promoting human resource development, technological innovation, and academic excellence (The University of Da Nang, n.d.-a).

### 4.6.1 AI Integration into Higher Education Curricula

UD promotes AI integration across its member institutions. For instance, the University of Science and Technology has established a center of excellence in AI and semiconductor education. This initiative is grounded in its strong industry partnerships, upgraded infrastructure including licensed AI software and laboratories and an expanding faculty (The University of Da Nang, 2025b).

Member universities of UD are incorporating AI into their curricula. For example, the University of Technical Education has implemented faculty development programs on AI integration. A June 2025 workshop introduced faculty and students to global AI practices, including tools such as ChatGPT, NotebookLM, Gemini, and Scite AI, with a focus on "safe, effective, and academically honest AI use" (The University of Da Nang, 2025c, para. 6). In the non-technical disciplines, the University of Foreign Language Studies organized a seminar on AI challenges and opportunities in language education in partnership with international experts from the University of Liège (The University of Da Nang, 2025a). The seminar fostered debates on the cultural and ethical factors of AI in education. It also encouraged AI use by students and faculty for language learning, research, and translation.

### 4.6.2 Alignment with Vietnam's AI Strategy and Policy Framework

The university's events and training initiatives aim to promote digitally responsible citizenship and prepare students for an AI-powered economy, echoing Vietnam's calls for ethical and human-centered AI education. Its member institution- the University of Science and Technology serves as a center for AI excellence indicates the UD's strong commitment to MOET's initiatives related to AI application (The University of Da Nang, 2025b). The institution is also training human capital in semiconductors and AI as a response to the country's strategy.

## *4.7 The Case of Vietnam National University, Hanoi*

Vietnam National University, Hanoi (VNU) has undertaken significant steps to integrate AI at the undergraduate and postgraduate levels.

### 4.7.1 AI Integration in Higher Education Curricula

VNU's higher education programs focus on digital transformation, digital citizenship, and AI thinking to equip students with essential twenty-first century competencies (VNU, 2025a). In addition, they aim to provide practical simulations and real-life case studies that support the application of AI knowledge beyond traditional STEM fields.

VNU's member institutions have incorporated AI into their academic programs. One example is that the University of Languages and International Studies (ULIS) organized a national training workshop on AI use in foreign language instruction and assessment for personalized feedback and automated evaluation in October 2023 (ULIS, 2023a, 2023b). This initiative was followed by a series of webinars in November 2025 that enabled lecturers across departments to share experiences and techniques for incorporating AI into classroom instruction and academic support services (ULIS, 2025).

Another example was provided by VNU's Information Technology Institute in 2024 when it expanded international collaborations, published AI-focused textbooks, and implemented AI-based tools for research and teaching (VNU, 2025b). These developments reflect VNU's commitment to enhancing AI capacity at the postgraduate level and contributing to Vietnam's growing AI research.

### 4.7.2 Alignment with Vietnam's AI Strategy and Policy Framework

VNU's AI-focused research, textbook publication, and professional development initiatives also support policy implementation related to capacity-building and innovation in teaching and learning. VNU embeds AI and digital skills into general education from the outset and contributes to the national goals of establishing foundational digital competency among students. Furthermore, VNU's member institution – the University of Economics and Business has supported national efforts through initiatives exploring how to integrate artificial and human intelligence for digital transformation (VNU, 2025c). Overall, VNU demonstrates its role as a leading university that can realize national AI policies in meaningful and context-specific ways.

## 4.8 *The Case of Vietnam National University-Ho Chi Minh City*

Established in 1995, Vietnam National University-Ho Chi Minh City (VNU-HCM) has been recognized as one of the most prestigious multidisciplinary institutions in Vietnam and for its innovation in science, technology, and social sciences ranks it as a national leader in educational reform (VNU-HCM, 2022). With over 90,000 full-time students enrolled across its eight member universities, VNU-HCM is advancing the country's higher education and research.

### 4.8.1 AI Integration into Higher Education Curricula

VNU-HCM has taken important measures to advance AI integration. It has invested more than 70 billion VND in specialized laboratories focusing on AI, intelligent systems, and information security (University of Science - VNU-HCM, n.d.). As one member institution, the University of Science (HCMUS) introduced the first degree programs in AI in southern Vietnam at both undergraduate and graduate levels with a curriculum covering foundational areas like machine learning and computer vision (University of Science - VNU-HCM, n.d.). These facilities enable students to work in small cohorts with experienced faculty and apply their learning through interdisciplinary research projects in fields such as healthcare, transportation, and finance. In addition, AI is being integrated into institutional governance and ethics policy. The University of Social Sciences and Humanities (USSH), another member of VNU-HCM, introduced strict academic integrity guidelines regarding AI use. The university's website states that researchers and writers are required to report the proportion of content supported by AI in their works, lectures, papers, and research (USSH - VNU-HCM, 2024, para. 5).

### 4.8.2 Alignment with Vietnam's AI Strategy and Policy Framework

VNU-HCM's AI integration is aligned with Vietnam's AI strategy by 2030 and MOET's initiatives. Since 2008, the university has operated one of Vietnam's earliest AI laboratories for research on machine learning, speech translation, and bioinformatics (Taipei Economic & Cultural Office in Vietnam, 2023). These efforts contribute to the national goals of strengthening AI research capacity and interdisciplinary innovation. In April 2024, VNU-HCM further advanced this commitment by entering a strategic partnership with Deakin University from Australia to cooperate in AI research and doctoral training. This collaboration realizes VNU-HCM's response to its 2021–2030 development strategy with AI playing a central role in advancing research and graduate education (University of Information Technology—VNU-HCM, 2024). Besides, the establishment of 13 key laboratories two of which operate at the state level as well as the university's AI-related publications and patents indicate VNU-HCM's contribution to national priorities in science, technology, and human capital development (Taipei Economic & Cultural Office in Vietnam, 2023).

## 5 Discussion

The findings of the study indicate that the eight universities in Vietnam have integrated AI into their higher education curricula. The first group includes VNU-Hanoi, RMIT, and UD, which demonstrates high degrees of AI integration in the core, interdisciplinary courses, and faculty training programs. CNU and CTU have made noticeable progress in this area through faculty workshops and selective course modules. The results resonate with the findings of Schreier (2012) who points out that institutional practice in policy implementation varies.

The case analysis also suggests that there must be balance in AI incorporation between automation and human-centered pedagogy as pointed out by Crompton and Burke (2023). It is evident in Hue University's medical ethics training and Fulbright University Vietnam's interdisciplinary AI courses. Moreover, TNU and Hue University embedded AI into curricula taking cultural and pedagogical traditions into account as shown in the study by Nguyen (2023).

William (2025) argues that successful AI adoptions requires institution and resource readiness as well as stakeholder engagement. The analysis of the current study shows that RMIT and Fulbright University Vietnam are better prepared for these shifts as shown in their international collaboration and policy implementation. Other universities are found to witness some challenges in infrastructure and expertise.

The study adopted Fullan's (2015) educational change framework to analyze how the key universities in Vietnam respond to Vietnam's AI strategy by 2030. The interplay of three factors of policy vision, institutional capacity, and professional agency is needed for the educational transformation to take place. In the study, VNU Hanoi and VNU-HCM demonstrate capacity building, AI curricula development, and faculty training, aligning with the country's policy framework. Fullan (2015) emphasizes collective efforts for educational transformation, which is reflected in the activities for international collaboration and interdisciplinary programs offered by RMIT and Fulbright University Vietnam in this study. This finding shows the interaction between the policy and institution for AI incorporation is needed for meaningful change. It also indicates that other institutions like CTU and TNU show early and adaptive changes within their resource constraints.

## 6 Conclusion and Contribution

The study presents an exploratory analysis of AI adoption among the eight universities in Vietnam. Adopting Fullan's (2015) educational change framework, the findings reveal that the reform requires not only technical expertise but also the interplay between capacity readiness, moral purpose, and collaborative professionalism. The eight universities adopt AI at different levels corresponding to the nations's strategy by 2030 and their capacity.

For the reform to take place, it is suggested that Vietnam's MOET provide guidelines for universities to incorporate AI in their curricula as well as sustained support for its realization. Furthermore, collaboration among the institutions is needed for AI knowledge and skill sharing for meaningful educational change. Future studies can further investigate other factors such as teacher and student readiness for AI integration ino higher education curricula via direct interviews with them.

**Acknowledgement** I thank ChatGPT for proofreading the manuscript.

## References

Alqahtani, T., Badreldin, H. A., Alrashed, M., Alshaya, A. I., Alghamdi, S. S., Bin Saleh, K., & Albekairy, A. M. (2023). The emergent role of artificial intelligence, natural learning processing, and large language models in higher education and research. *Research in Social and Administrative Pharmacy, 19*(8), 1236–1242. https://doi.org/10.1016/j.sapharm.2023.05.016

Bui, T. T., Phi, D. K., & Pham, T. T. H. (2024). Nghiên cứu nhận thức của sinh viên Trường Đại học Khoa học–Đại học Thái Nguyên về sử dụng trí tuệ nhân tạo trong hoạt động học tập. *Tạp chí Giáo dục*, 319–325. https://tcgd.tapchigiaoduc.edu.vn/index.php/tapchi/article/view/2241

Can Tho University. (n.d.). *General introduction.* https://en.ctu.edu.vn/introduction/general.html

Can Tho University. (2025, June 15). *Trường Đại học Cần Thơ nắm bắt xu thế, nâng cao năng lực AI trong giáo dục đại học* [CTU advances AI capacity in higher education]. https://www.ctu.edu.vn/tin-tuc-su-kien-t/truong-dai-hoc-can-tho-nam-bat-xu-the-nang-cao-nang-luc-ai-trong-giao-duc-dai-hoc.html

Chan, C. K. Y., & Tsi, L. H. (2023). The AI revolution in education: Will AI replace or assist teachers in higher education? *arXiv preprint*, https://doi.org/10.48550/arXiv.2305.01185

College of Agriculture—Can Tho University. (2025). *Thông tin seminar: Sử dụng AI trong viết báo khoa học* [Seminar on AI in academic writing]. https://coa.ctu.edu.vn/nghien-cuu/seminar-hoc-thuat/1810-thong-tin-seminar-su-dung-ai-trong-viet-bao-khoa-hoc.html

Crompton, H., & Burke, D. (2023). Artificial intelligence in higher education: The state of the field. *International Journal of Educational Technology in Higher Education, 20*(1), 1–22. https://doi.org/10.1186/s41239-023-00409-7

Donnell, F. O., Porter, M., & Fitzgerald, S. (2024). The role of artificial intelligence in higher education: Higher education students' use of AI in academic assignments. *Irish Journal of Technology Enhanced Learning, 8*(1), 1–15. https://doi.org/10.22554/ijtel.v8i1.155

FPT & Hue University of Medicine and Pharmacy. (2024, May 17). *FPT and Hue University of Medicine and Pharmacy join hands to advance the digital transformation of education and healthcare.* https://fpt.com/en/news/fpt-news/fpt-hop-tac-dh-y-duoc-hue

Fulbright University Vietnam. (n.d.-a). *Mission and vision.* https://fulbright.edu.vn/our-mission/

Fulbright University Vietnam. (2023, July). *AI and the future of public policy: A student perspective.* https://fulbright.edu.vn/ai-and-the-future-of-public-policy/
Fulbright University Vietnam. (2023, September). *Artificial intelligence meets the liberal arts.* https://fulbright.edu.vn/wp-content/uploads/2024/01/Fulbright-Press-Release-Artificial-Intelligence-meets-the-Liberal-Arts-12-September-20231.pdf
Fulbright University Vietnam. (n.d. b). Academics. https://fulbright.edu.vn/academics/
Fulbright University Vietnam (2024). Class of the fearless. https://fulbright.edu.vn/commencement-2024/
Fullan, M. (2015). *The new meaning of educational change.* Teachers college press.
Giao Duc Thoi Dai. (2025, April 26). *Làm chủ AI, nâng tầm kỹ năng số cho sinh viên Đại học Huế* [Mastering AI and enhancing digital skills for Hue University students]. https://giaoducthoidai.vn/lam-chu-ai-nang-tam-ky-nang-so-cho-sinh-vien-dai-hoc-hue-post728839.html
Government of Vietnam. (2021). *National strategy on research, development and application of artificial intelligence until 2030.* https://en.baochinhphu.vn/national-strategy-on-rd-and-application-of-artificial-intelligence-11140663.htm
Hue University. (2020, March 6). *Tuyển sinh 2020: Khoa học dữ liệu và Trí tuệ nhân tạo – Phân biệt sự khác nhau giữa hệ kỹ sư và hệ cử nhân.* [Admissions 2020: Data science and artificial intelligence – Distinguishing the differences between the engineering and bachelor's degree programs.] https://hueuni.edu.vn/portal/vi/index.php/News/tuyen-sinh-2020:-khoa-hoc-du-lieu-va-tri-tue-nhan-tao-phan-biet-su-khac-nhau-giua-he-ky-su-va-he-cu-nhan.html
Hue University. (n.d.-a). *Development history of Hue University.* https://hueuni.edu.vn/portal/en/index.php/News/development-history-of-hue-university.html
Hue University. (n.d.-b). *Mission and vision.* https://hueuni.edu.vn/portal/en/index.php/Category/mission-and-vision.html
HueMed. (2025, May 17). *Science Café: Chuyển đổi số và ứng dụng trí tuệ nhân tạo (AI) trong Y khoa.* [Science Café: Digital transformation and the application of artificial intelligence (AI) in medicine]. https://huemed-univ.edu.vn/tin-tuc/hoat-dong-hue-ump/science-cafe--chuyen-doi-so-va-ung-dung-tri-tue-nhan-tao--ai--trong-y-khoa-3483
Lan, D. T. H., & Van, D. T. (2025). The challenges of using AI in studying for students at School of Foreign Languages—Thai Nguyen University—Vietnam. *International Journal of All Research Writings, 6*(9), 142–148.
Nguyen, N. D. (2023). Exploring the role of AI in education. *London Journal of Social Sciences, 6*, 84–95.
Quy, V. K., Thanh, B. T., Chehri, A., Linh, D. M., & Tuan, D. A. (2023). AI and digital transformation in higher education: Vision and approach of a specific university in Vietnam. *Sustainability, 15*(14), 11093.
RMIT Vietnam. (n.d.-b). *Artificial intelligence study area.* https://www.rmit.edu.vn/study-at-rmit/study-areas/artificial-intelligence
RMIT Vietnam. (n.d.-a). *About us.* https://www.rmit.edu.vn/about-us
RMIT Vietnam. (2020, March 3). *Vietnam getting ready to ride Artificial Intelligence wave.* https://www.rmit.edu.vn/news/all-news/2020/mar/vietnam-getting-ready-to-ride-artificial-intelligence-wave
RMIT Vietnam. (n.d.-c). *Master of Artificial Intelligence.* https://www.rmit.edu.vn/study-at-rmit/postgraduate-programs/master-of-artificial-intelligence
Schreier, M. (2012). *Qualitative content analysis in practice.* SAGE Publications.
Taipei Economic and Cultural Office in Vietnam. (2023). *Vietnam National University-HCM City, a pioneer in AI research.* https://www.roc-taiwan.org/vn_en/post/2905.html
Thai Nguyen University. (n.d.). *Lịch sử phát triển* [Development history]. https://tnu.edu.vn/gioi-thieu/lich-su-va-phat-trien/lich-su-phat-trien
Thai Nguyen University. (2023, November 10). *Đại học Thái Nguyên tập huấn về ứng dụng công nghệ trí tuệ nhân tạo AI cho giảng viên* [Thai Nguyen University conducts training on the application of artificial intelligence (AI) technology for

lecturers]. https://tnu.edu.vn/tin-tuc-su-kien/dai-hoc-thai-nguyen-tap-huan-ve-ung-dung-cong-nghe-tri-tue-nhan-tao-ai-cho-giang-vien2.html

Thai Nguyen University. (2024, April 23). *Hơn 300 cán bộ giảng viên dự tập huấn ứng dụng trí tuệ nhân tạo.* [More than 300 faculty members attended the training on artificial intelligence application.] https://tnu.edu.vn/tin-dai-hoc-thai-nguyen/hon-300-can-bo-giang-vien-du-tap-huan-ung-dung-tri-tue-nhan-tao.html

The University of Da Nang. (n.d.-a). *History of formation.* https://www.udn.vn/english/about-us/history-of-formation

The University of Da Nang. (2025a, June 22). *Kết nối chuyên gia thúc đẩy ứng dụng ngoại ngữ và trí tuệ nhân tạo trên giảng đường* [Connecting experts to promote AI and foreign language application in the classroom]. https://udn.vn/tin-tuc/chi-tiet/ket-noi-chuyen-gia-thuc-day-ung-dung-ngoai-ngu-va-tri-tue-nhan-tao-tren-giang-duong

The University of Da Nang. (2025b, June 17). *Trường Đại học Bách khoa - ĐHĐN được chọn tham gia Đề án phát triển trung tâm đào tạo xuất sắc và tài năng về trí tuệ nhân tạo và vi mạch bán dẫn* [University of Science and Technology – UD selected for AI center of excellence]. https://www.udn.vn/tin-tuc/chi-tiet/truong-dai-hoc-bach-khoa-dhdn-duoc-chon-tham-gia-de-an-phat-trien-trung-tam-dao-tao-xuat-sac-va-tai-nang-ve-tri-tue-nhan-tao-va-vi-mach-ban-dan

The University of Da Nang. (2025c, June 23). *Trường Đại học Sư phạm Kỹ thuật - ĐHĐN tập huấn ứng dụng AI trong đào tạo và nghiên cứu* [University of Technical Education – UD trains in AI for teaching and research]. https://www.udn.vn/tin-tuc/tin-tuc-su-kien/chi-tiet/truong-dai-hoc-su-pham-ky-thuat-dhdn-tap-huan-ung-dung-ai-trong-dao-tao-va-nghien-cuu

ULIS (2023, October). *Hội thảo tập huấn: "Ứng dụng công nghệ trí tuệ nhân tạo (AI) trong dạy, học và kiểm tra, đánh giá ngoại ngữ*". https://ulis.vnu.edu.vn/hoi-thao-tap-huan-ung-dung-cong-nghe-tri-tue-nhan-tao-ai-trong-day-hoc-va-kiem-tra-danh-gia-ngoai-ngu/

ULIS. (2023, October). Hội thảo tập huấn: *Ứng dụng công nghệ trí tuệ nhân tạo (AI) trong dạy, học và kiểm tra, đánh giá ngoại ngữ*.[ Training workshop: Application of artificial intelligence (AI) technology in foreign language teaching, learning, and assessment]. https://ulis.vnu.edu.vn/hoi-thao-tap-huan-ung-dung-cong-nghe-tri-tue-nhan-tao-ai-trong-day-hoc-va-kiem-tra-danh-gia-ngoai-ngu/

ULIS. (2025, November). *Triển khai chuỗi Webinar chia sẻ kinh nghiệm ứng dụng AI trong công tác giảng dạy và phục vụ đào tạo tại ULIS* [Implementing a series of webinars to share experiences in applying AI in teaching and educational support at ULIS.] https://ulis.vnu.edu.vn/trien-khai-chuoi-webinar-chia-se-kinh-nghiem-ung-dung-ai-trong-cong-tac-giang-day-va-phuc-vu-dao-tao-tai-ulis/

University of Information Technology—VNU-HCM. (2024, April 8). *VNU-HCM and Deakin University signed strategic cooperation agreement on AI research and education.* https://en.uit.edu.vn/vnu-hcm-and-deakin-university-signed-strategic-cooperation-agreement-ai-research-and-education

University of Sciences—Hue University. (2025, May 16). *New learning styles—Leading the trend with AI*. http://husc.hueuni.edu.vn/en/news.php?readmore=1885

University of Science—VNU-HCM. (n.d.). *Artificial intelligence undergraduate program.* https://www.fit.hcmus.edu.vn/undergraduate/artificial-intelligence

University of Social Sciences and Humanities—VNU-HCM. (2024, January 15). *Member schools of Ho Chi Minh City National University have strict regulations when using artificial intelligence (AI).* https://www.vietnam.vn/en/truong-thanh-vien-dh-quoc-gia-tp-hcm-quy-dinh-liem-chinh-hoc-thuat-khi-dung-ai

Vietnam.vn. (2024, November). *Can Tho University opens many new majors, including Artificial Intelligence.* https://www.vietnam.vn/en/truong-dai-hoc-can-tho-mo-them-nhieu-nganh-moi-trong-do-co-tri-tue-nhan-tao

VNU. (2025c, June). *Tích hợp trí tuệ nhân tạo với trí tuệ con người trong chuyển đổi số trên thế giới.* https://vnu.edu.vn/tich-hop-tri-tue-nhan-tao-voi-tri-tue-con-nguoi-trong-chuyen-doi-so-tren-the-gioi-post38323.html

VNU (2025, May). *ĐHQGHN đưa học phần công nghệ số và trí tuệ nhân tạo vào giảng dạy từ năm nhất.* https://vnu.edu.vn/dhqghn-dua-hoc-phan-cong-nghe-so-va-tri-tue-nhan-tao-vao-giang-day-tu-nam-nhat-post36665.html

VNU. (2025b, February). *10 Key events and achievements of the information technology institute in 2024.* https://iti.vnu.edu.vn/en/10-key-events-and-achievements-of-the-information-technology-institute-in-2024/

VNU. (2025a, May). *ĐHQGHN đưa học phần công nghệ số và trí tuệ nhân tạo vào giảng dạy từ năm nhất.* [VNU has introduced digital technology and artificial intelligence courses into the curriculum starting from the first year].

VNU-HCM. (2022, December 21). *About us.* https://vnuhcm.edu.vn/about-us/33393364/303364/303364

Vocal Media. (2024, June). *Exploring the frontier: Artificial intelligence research initiatives and outcomes in Can Tho, Vietnam.* https://vocal.media/education/exploring-the-frontier-artificial-intelligence-research-initiatives-and-outcomes-in-can-tho-vietnam

Williams, A. (2025). Integrating artificial intelligence into higher education assessment. *Intersection: A Journal at the Intersection of Assessment and Learning*, *6*(1).

Yin, R. K. (2018). *Case study research and applications: Design and methods* (6th ed.). SAGE Publications.

**Le Pham Hoai Huong** is an associate professor in the English Department at Hue University of Foreign Languages and International Studies, Hue University, Vietnam. Her research interests include sociocultural theory, tool-mediated language learning, and language education curriculum development.

# AI Applications in Teaching and Learning Contexts

# Turkish Students' Use of NotebookLM in an Online Digital Literacy Course

**Bünyami Kayali and Mehmet Yavuz**

**Abstract** This experimental study investigates the effects of using NotebookLM on students' academic achievement and students' assessment of NotebookLM in online digital literacy courses in higher education. It was conducted using an explanatory sequential mixed-methods design. A one-group pretest–posttest model was employed. During a five-week intervention, 33 Turkish students used NotebookLM for summarizing, concept mapping, and scenario-based problem-solving activities. Data was collected from a pretest, posttest, and interviews. Quantitative data was projected into SPSS 27 for analysis, and qualitative data was thematically analysed. Results reveal a significant difference between the pretest score (M = 66.90, SD = 10.26) and posttest score (M = 74.90, SD = 13.16) in terms of academic achievement, $t(32) = -3.51$, $p = 0.001$. The effect size was moderate (Cohen's $d = 0.68$). Interview data shows that the students perceived NotebookLM as helpful for comprehension, time efficiency, motivation, and feedback, though constrained by language support and occasional technical problems. The findings suggest implications for using NotebookLM in online courses in higher education in Turkey.

**Keywords** Artificial intelligence · AI in education · NotebookLM · Online learning · Digital literacy

## 1 Introduction

The dizzying speed of digitalisation has affected all areas of social life, including education. The intensive use of digital tools by higher education institutions and the subsequent COVID-19 pandemic have significantly transformed education and

B. Kayali
Technical Sciences Vocational School, Bayburt University, Bayburt, Türkiye

M. Yavuz (✉)
Distance Education Application and Research Center, Bingöl University, Bingöl, Türkiye
e-mail: myavuz@bingol.edu.tr

 
H. P. Bui and L. P. H. Huong (eds.), *Artificial Intelligence in Education and Academic Research*, Studies in Computational Intelligence 1269,
https://doi.org/10.1007/978-3-032-18392-7_3

supported the spread of online learning activities (Bui et al., 2023). E-learning activities, which have already caught an upward trend with the effect of the digitalisation process, have become a fundamental component of education systems in a short time and have become the "new normal" (Chukwuere, 2024).

Online learning offers flexibility in time and place and can be cost-efficient through reusable content and reduced materials, supporting self-paced, independent study and richer media use; students also value online testing and clear criteria (Cohen & Nycz, 2006; Goyal, 2012; Jethro et al., 2012; Kuimova et al., 2016; Radović-Marković, 2010; Yusuf & Al-Banawi, 2013). Yet key challenges persist reduced face-to-face interaction and collaboration, difficulties with practical subjects and motivation, limited teacher preparation and integration skills, and technical constraints such as devices, connectivity, and infrastructure dense content can also strain cognitive processing (Klisowska et al., 2020; Preniqi et al., 2020; Shet, 2020; Stecuła & Wolniak, 2022).

The use of AI technology such as large language models (LLMs) may enhance online learning by providing specific feedback that promotes summarisation, creation of concept maps, engagement and learning efficiency (Holmes et al., 2019; Hwang et al., 2022; Luckin et al., 2016). Despite the advantages of AI for educational environments, various studies also highlight the negative aspects of this technology. Studies show that overuse of or overdependence on AI tools can weaken students' creativity and learning autonomy (Rabadanova et al., 2024). Similarly, excessive reliance on artificial intelligence may lead to the gradual erosion of fundamental critical thinking over time (Ivanov, 2023). It is also noted that these technologies can reduce social interaction in educational settings and negatively limit the social role of teachers in classrooms (Chisega-Negrilă, 2024). Finally, studies also emphasize ethical and practical risks, such as algorithmic bias, plagiarism, loss of authenticity, and data privacy (Ivanov, 2023).

LLM-based AI tools suggest being facilitative for learning. These tools can enable students to create summaries, generate question–answer pairs, and develop personalized learning guides based on the course content uploaded by students. These possibilities contribute to students not only accessing information but also interacting with the content in an active and personalised way. The NotebookLM tool, developed by Google, stands out as a remarkable and important example. NotebookLM can summarize the lecture notes uploaded by students, provide explanations to enhance understanding of the subject, answer questions about the content, and produce various types of content, such as podcasts. Thanks to these features, students' learning autonomy is supported, and students with different learning styles are supported. In this way, it is thought to contribute to the solution of problems such as high cognitive load and learning difficulties caused by different content types, especially in online learning environments. However, while existing research generally evaluates such tools on technical accuracy and information reliability, it does not focus enough on their effects on student experiences and learning outcomes (Swacha & Gracel, 2025). The present study aims to examine students' experiences of utilising NotebookLM in an online digital literacy course. It focuses on investigating the impact of these experiences on their perspectives. Furthermore, the impact of students' experiences using NotebookLM in an online digital literacy course on their academic achievement was examined. In this study, answers were sought to the following research questions.

RQ1. What is the effect of using NotebookLM on Turkish L2 English students' academic achievement in an online digital literacy course?

RQ2. How do Turkish L2 English students assess the use of NotebookLM in the overall learning experience?

## 2 Literature Review

### *2.1 AI in Education*

AI is the general name for systems that can simulate complex cognitive processes in the human mind, such as such as acquiring knowledge and addressing complex tasks (McCarthy, 2007). In recent years, it has initiated a personalized learning experience, tracked process-based student development, and provided administrators with data-based decision-making for providing personalized learning experiences, tracking process-based student development, and enabling administrators to make data-driven decisions (Luckin et al., 2016). With the COVID-19 pandemic, technology-based educational tools have been rapidly adopted, and AI applications have become increasingly visible in educational environments. Therefore, artificial intelligence is described as a technology that supports teachers and improves learning processes (Alanoğlu & Karabatak, 2020).

However, the current literature acknowledges some limitations and ethical challenges related to the application of AI in educational settings (Vincent-Lancrin & Van der Vlies, 2020). Experts state that AI should not be allowed to direct the educational process completely. Instead, it is recommended to use it under the guidance of decision makers and educators, adhering to pedagogical principles (Zawacki-Richter et al., 2019). Therefore, it is necessary to correctly define AI, determine its boundaries, and use it effectively and responsibly. Otherwise, the technological tools used to achieve educational goals run the risk of being reduced to a mere technical tool by moving away from these missions (Alanoğlu & Karabatak, 2020).

## 2.2 AI in Open and Distance Education

AI holds an important place in open and distance learning (ODL) environments thanks to its many contributions (Hwang et al., 2022). These contributions include customisable e-learning platforms, process-based student tracking and data-driven predictions. Especially in massive open online courses (MOOCs) with high student participation, one of the most important problems is the inability to track the students participating in the course. In these cases, AI tools have become an important tool to facilitate the follow-up of students and to improve participation and success rates. The main applications of AI in ODL include student tracking, success prediction, automatic assessment and personalised learning (Özkan, 2019).

However, the intensive use of AI in educational activities raises various issues and ethical concerns. Questions about data privacy, algorithmic biases, the use of costly systems, and transparency throughout decision processes are at the centre of these concerns. Therefore, although AI technologies have the potential to offer important opportunities for ODL, the ethical, pedagogical, and technical dimensions of their use need to be carefully examined. Otherwise, it may lead to a deepening of inequalities in educational opportunities, while rights violations increase (Şenocak, 2020).

## 2.3 LLM and NotebookLM

LLMs are models built on an artificial neural network architecture called transformer and trained with the help of books, journals, and online forums, among other sources of diverse textual data (Filippi & Motyl, 2024). By using deep learning algorithms, these models can reveal complex patterns of human language and thus produce human-like, contextually relevant, and meaningful texts (Shahzad et al., 2025). For example, ChatGPT is one of the most advanced and well-known examples among LLMs. Thanks to the extensive pre-training and subsequent fine-tuning processes mentioned above, these language models can generate content, summarise, translate, and answer questions in natural language processing tasks.

In recent years, LLMs have found a wide place in educational settings. These models contribute to the efficiency of teaching services by automating processes such as content generation, assessment activities and designing personalised learning environments (Wollny et al., 2021). In this context, educators prefer these models to prepare course content, automate assessment activities or provide personalised feedback to students. Students, on the other hand, prefer to use these technologies to understand complex concepts, summarise long texts, or request annotations. Researchers frequently emphasise the potential of LLMs to encourage creativity, increase access and enhance learning (Ülkü, 2023). These applications play a supportive role in educational activities and contribute to improving their quality (Peláez-Sánchez et al., 2024).

As a result, although it provides essential findings on the general contributions and challenges of AI tools, research directly examining the effects of LLM-based and student-centred tools such as NotebookLM on student experiences and learning outcomes in online courses is quite limited (Alisoy, 2025; Miranda, 2025). Most studies have focused on the efficiency of generative AI (Ojha & Kadel, 2024), but the effects of specific applications, such as summarization, podcast production, or content transformation, on students' cognitive processes have not been examined. The study aims to address this gap identified in the previous literature. Investigating the effects of NotebookLM on student experiences, particularly in dimensions such as motivation, cognitive load, self-organized learning, and academic performance, will both contribute to theoretical literature and provide pedagogical guidance for practitioners.

# 3 Research Methodology

## *3.1 Research Design*

An explanatory sequential mixed-methods design was deployed. In the current study, quantitative data was initially collected, and then qualitative data was used to examine the quantitative findings in more depth (Creswell, 2021). To examine the effect of NotebookLM on academic achievement, a one-group pretest–posttest quasi-experimental design was used. The design tests the effect of the intervention on only one group (Büyüköztürk et al., 2018). Then, semi-structured interviews were conducted to understand students' assessment of NotebookLM use in a detailed and systematic way (Berg & Lune, 2012).

## *3.2 Participants*

Thirty-three students currently enrolled at the Department of Management Information Systems of a university in Türkiye participated in the study. They were selected using convenience sampling. Due to accessibility and voluntary participation constraints in the online environment, convenience sampling was preferred. An informed consent form has been prepared for potential students. The form contains detailed information about the process, such as the purpose of the study, its duration, and the data collection tools. In addition to the consent form, verbal announcements were made in the classrooms, and details about the study were shared with the potential students. Students participating in the voluntary work programme were asked to communicate either in person or via email. At the end of the process, 33 students decided to take part in the intervention phase. However, four participants did not agree to participate in the interview phase, resulting in only 29 students in the interviews. They were anonymized as S1-S29 in the data report.

## 3.3 Data Collection

This study used two different data collection tools. First, a pretest and posttest were administered before and after the application to measure the effects of NotebookLM use. The pretest and posttest (see Appendix B) were designed to measure students' academic achievement in the online digital literacy course, with a focus on their comprehension, analysis and application of the course content before and after using NotebookLM. Initially, a pool of 30 items was created to develop the tests. Two field experts were invited to analyse these tools. As a result, five items were removed as suggested by the experts. The remaining 25 items were piloted to seven students not involved in the main study. After the pilot application, the form was reviewed by a language expert for language accuracy. After this pilot testing phase, the tests consisted of 25 items each. In the analysis of the academic achievement test, item difficulty and discrimination coefficients were determined. For the final version of the 25-item tests, the internal consistency (Cronbach's alpha) was calculated as $\alpha = 0.86$. This value indicates a high scale reliability.

Qualitative data obtained through semi-structured interviews. The interview questions were designed to determine students' experiences of using NotebookLM. For this purpose, the interview scheme (see Appendix A) mainly focused on the purpose of using the model, its advantages, difficulties experienced in its use, interaction, and suggestions for improvement. We applied the guidelines for interviews by Khoa et al. (2023) and Bui and Vo (2024). A total of 13 questions were initially developed from an extensive literature review. It was then reviewed by two experts and one Turkish language expert. Following the necessary corrections, the interview questions were piloted with three students. The interview questions were finalised based on the data obtained from the pilot stage. All interviews were recorded, with student content, to prevent data loss.

## 3.4 Data Analysis

Descriptive and predictive statistical techniques were employed to analyse the quantitative data. The analyses were carried out using SPSS 27 software. In this context, the aim was to determine the change in scores from the pretest to the posttest. First, the Shapiro–Wilk normality test was performed to examine the distribution of the data. The results showed that the distribution was normal ($Zpre = 0.61$; $Zpost = 0.42$; $p > 0.05$). Therefore, we decided to use parametric tests. A paired samples t-test was applied to compare the pretest and posttest scores to investigate the effect of the intervention.

The interview data was analysed thematically. To ensure confidentiality, all student participants were given pseudonyms (S1 – S29). Initially, the recordings were transferred to the computer environment and converted into written texts. The transcripts were analysed in three phases. First, the data was compiled. Second, the data were coded, and a code list was created consistent with the study's objective. To verify the consistency of the coding, the coding process was reviewed again, and similar codes were consolidated. In the last stage, the codes serving the same purpose were categorized according to their common characteristics, and themes were obtained using these categories. The results obtained after the analysis were interpreted, visualised and reported in detail. The reliability and credibility of the qualitative data were verified through triangulation and peer review. The transcripts of the interviews were analysed by two different coders. Finally, to ensure the reliability of the interpretations, the transcripts were reviewed by the participants for participant confirmation. The coding system of the interview data is presented in Table 1.

## *3.5 Research Process*

This study investigated students' use of NotebookLM in an online digital literacy course conducted for five consecutive weeks using an explanatory sequential mixed method. A one-group pretest–posttest design was implemented: students completed a pretest, engaged in structured NotebookLM-supported activities each week (e.g., summarizing source materials, concept mapping, and scenario-based problem solving) under instructor guidance, and then completed a posttest. In the qualitative phase, a subset of students took part in semi-structured interviews to elaborate on their experiences with the tool and to contextualize the quantitative results. Quantitative analyses compared pre- and post-intervention academic achievement, while qualitative content analysis was used to identify perceived benefits and challenges. Table 2 summarizes the week-by-week sequence of activities and assessments.

According to Table 2, each week of the course was structured by focusing on a different function of the NotebookLM tool. In the first week, the introduction and exploration process enabled students to experience the basic features, while the second week emphasized the ability to construct knowledge. In the third week, critical thinking was supported through comparison and analysis activities. In the fourth week, digital content production was developed through problem-solving and scenario-based learning. In the fifth and final week, students are expected to develop creative products using what they learned and practiced in the previous weeks.

**Table 1** Coding system (n = 29 interviews)

| Theme | Category | Codes | f |
|---|---|---|---|
| Contribution to the learning process | Facilitating understanding | Better comprehension of the subject, simplification of complex concepts, faster understanding, increased focus, and reduction of concept confusion | 52 |
| | Summarization and simplification | Summarizing long texts, presenting short and precise information, simplifying lecture notes, ease of repetition, and organizing information | 61 |
| | Time saving | Quick access, shortening repetitions, learning in a short time, speeding up studying, and practical use | 78 |
| | Increasing motivation | Making studying enjoyable, encouraging learning, arousing more curiosity, developing self-confidence, giving a sense of academic achievement | 44 |
| Technical experience | Ease of use | Simple interface, intuitive operation, user-friendly design, easy access, fast loading | 47 |
| | Technical challenges | No problems, smooth operation of the system, ease of access, low error rate, and sometimes connection problems | 40 |
| Feedback and support | Contribution to learning | Recognizing mistakes, reinforcing correct learning, providing immediate feedback, guiding explanations, increasing learning speed, and establishing interaction | 46 |
| | Teacher interaction | Alignment with teacher grades, facilitating communication with the teacher, supporting the given content, easing the teacher's burden, ease of co-operation | 43 |
| Suggestions and improvement requests | Language and access | Full Turkish support, more advanced language processing, translation options, read aloud, mobile compatibility | 35 |
| | New features | Adding images, graphic/chart support, interactive modules, sharing with friends, suggesting a personal study plan, adding quiz/test | 39 |
| Intent to use | Future use | Use in verbal courses, useful for history/literature, preference in courses with dense content, use in exam preparation, use for review purposes | 50 |
| | Advice to others | Recommendation to friends, suggestion to other students, finding practical solutions, increasing efficiency, and contributing to academic success | 42 |

**Table 2** NotebookLM activities (n = 33)

| Week | Purpose | Activities | Usage | Outcomes |
|---|---|---|---|---|
| *Week 1* Introduction and exploration | Recognizing the tool, discovering basic functions | An article titled "What is digital literacy?" has been uploaded. Students are asked to summarize the text, create a concept list, and produce audio content using the Podcast feature | Summarizing, concept extraction, and Podcast production | Recognize basic functions, experience content in different formats |
| *Week 2* Structuring knowledge | Summarizing and concept mapping from different sources | Notes on "Internet Technologies and Digital Infrastructure" are uploaded. After receiving the summaries, students are asked to visualize the concept relationships with the mind map feature of NotebookLM | Summarizing, mind mapping | Structuring knowledge, establishing relationships between concepts |
| *Week 3* Comparison and analysis | Similarity-difference analysis between sources | The report on "Social Media and Digital Communication" is uploaded. Comparison is taken from NotebookLM. Then, using the "question–answer" feature of the system, students are asked questions about the texts | Comparison, question–answer, analysis | Critical thinking, making connections between texts |

(continued)

**Table 2** (continued)

| Week | Purpose | Activities | Usage | Outcomes |
|---|---|---|---|---|
| *Week 4* Problem solving | Scenario-based learning | A problem scenario on "digital security" is given. Students collect resources, get solution suggestions with NotebookLM, and then visualize the solution scenario with the video generation feature | Source synthesis, solution generation, video creation | Problem-solving, digital content production |
| *Week 5* Product development | Transforming knowledge into creative products | In week 4, students extend the solution they developed and prepare a report, podcast, video, or infographic. They receive content streaming and language support from NotebookLM | Conversion of outputs, multi-format product development | Creativity, productivity, and digital presentation skills |

# 4 Findings

## 4.1 The Effect of Using NotebookLM on Students' Academic Achievement in an Online Digital Literacy Course

This title focuses on how the use of NotebookLM in the online digital literacy course affects students' academic achievement (see Table 3). Accordingly, male and female students' pretest scores were compared. Independent samples t-test was employed to examine the difference in scores between the two groups. Statistical analyses revealed no significant difference in academic achievement scores between female students ($N = 17$, $M = 67.06$, $SD = 11.88$) and male students ($N = 16$, $M = 66.75$, $SD = 8.61$), $t(31) = 0.09$, $p = 0.93$.

An independent sample t-test was conducted to elaborate the findings and to reveal the change in students' performance according to the gender factor. The results

**Table 3** Pretest independent samples t-test results according to gender

| Gender | N | *M* | *SD* | df | *t* | F | *p* |
|---|---|---|---|---|---|---|---|
| Female | 17 | 67.05 | 11.87 | 31 | 0.09 | 1.67 | 0.93 |
| Male | 16 | 66.75 | 8.60 | | | | |

**Table 4** Dependent groups t-test results of pretest and posttest scores

| Test | N | M | SD | df | *t* | *p* | Cohen's *d* |
|---|---|---|---|---|---|---|---|
| Pretest | 33 | 66.90 | 10.26 | 32 | −3.51 | 0.001 | 0.68 |
| Posttest | 33 | 74.90 | 13.16 | | | | |

showed that there was no significant difference between female (N = 17, M = 11.29, SD = 13.13) and male students (N = 16, M = 4.50, SD = 12.46), $t(31) = 1.52, p = 0.13$. Finally, dependent groups were examined using a t-test to assess the difference between the groups' pretest and posttest scores (see Table 4). Accordingly, a significant improvement was found between the students' pretest (M = 66.90, SD = 10.26) and posttest (M = 74.90, SD = 13.16) scores, $t(32) = -3.51, p = 0.001$. These results indicate that the NotebookLM tool had a positive effect on student academic achievement. Furthermore, according to Cohen's (1988) classification, a Cohen's *d* value of 0.68 indicates a medium effect size. This moderate effect size indicates that a meaningful pedagogical effect can be achieved within a relatively short intervention period.

## 4.2 Students' Experiences Regarding the Use of NotebookLM

In this section, students' opinions about the NotebookLM application used in the online digital literacy course are discussed. Student feedback has been grouped under five main themes. Students reported that NotebookLM facilitates comprehension, summarises and simplifies content, saves time, and increases motivation to learn. In the technical experience theme, the user-friendly structure of the application came to the forefront, and that technical problems were minimal. In the feedback and support dimension, instant feedback, recognizing mistakes, and the complementary role of teacher support are noteworthy. The students suggested that Turkish language support should be provided, more visualisation options should be added, and interactive modules should be incorporated. Finally, the students emphasised that they would use the application again to prepare for the exam, to repeat the lessons and to be successful in verbal lessons, and that they would advise their peers to use it as well.

Students also stated that the use of NotebookLM in their learning processes made significant contributions. Most students noted that the platform made the topics more understandable and reduced concept confusion (f = 52). Moreover, summarizing long texts and presenting short, clear information made it easier to repeat the lessons (f = 61). It is also seen that the platform saves time (f = 78) and accelerates the learning process. Some students emphasized that these facilities increased their motivation and made studying more enjoyable (f = 44). Some students reported:

> *It made it very easy for me to understand the subject; it saved me from reading long texts and summarized it helped me to grasp the subject faster, preventing confusion and reinforcing it because it provided voice narration, and I understood the subject well.* (S5)

*I used it to understand my lesson better, and the features I liked were that when I was having difficulty understanding long texts, when I used this application, it gave me a summary of the subject, which helped me to understand and learn better.* (S8)

Most students found NotebookLM easy and practical to use (f = 47). The intuitive and straightforward interface was evaluated positively by the students. It is seen that the number of students experiencing technical problems is quite limited, and the majority did not encounter any problems (f = 40). This indicates that the platform operates smoothly without causing significant technical issues. Some students said:

*When using NotebookLM for the first time, I was immediately struck by the simple and guidance-oriented interface. The file upload steps went smoothly, and the document content was processed quickly.* (S3)

*I have observed that the system sometimes slows down or gives errors, especially when loading large or complexly formatted PDF files.* (S12)

The feedback provided by NotebookLM was found to be supportive of the learning process by the students (n = 46). The students stated that they realized their mistakes more easily and that the feedback was helpful in terms of reinforcing correct learning. Additionally, it was noted that the platform works in harmony with teacher notes (f = 43) to facilitate communication and contribute to the teaching process. Two students stated:

*I think it was sufficient feedback. It contributed to my learning in every sense. (S1)*

*It was a more permanent learning for me in a more focused way, without a pile of unnecessary information, it was advantageous in many ways to read only the places that are important to me by reading the places that are important to me.* (S7)

Students made various suggestions for more efficient use of NotebookLM. The most frequently mentioned suggestion was to improve the platform's Turkish language support (f = 35). In addition, suggestions included adding visuals and graphics, developing interactive modules, and integrating quiz and test features (f = 39). Two students' responded:

*The platform's ability to fully understand the content in complex or tabular documents needs to be improved.* (S11)

*Slowdowns and delays, especially in large files, should be reduced. Makes it difficult to use.* (S21)

Most students stated that they would use NotebookLM again in the future (n = 50). It was noted that this platform would be especially useful in verbal courses, subjects with dense content, and exam preparation processes. In addition, a significant number of the students stated that they recommended NotebookLM to their peers (f = 42). They stated that they thought it would be useful for them to use such a tool. A few students answered:

*The most important contribution is that it saves time in preparing for exams, especially in courses with high course content.* (S13)

*I recommend the application to my other friends to use it. It makes studying, especially verbal lessons, more enjoyable.* (S23)

*I plan to use it in verbal courses such as history and literature where long texts are involved.* (S19)

## 5 Discussion

The findings about students' use of NotebookLM within the scope of the online digital literacy course conducted online are interpreted in parallel with the research questions by associating them with previous literature. The quantitative findings indicate that the use of NotebookLM results in a significant improvement in students' academic achievement. Qualitative data suggest that students have a positive perspective towards this technology due to its contributions, such as producing digital content easily, increasing motivation, and playing a supportive role. However, some students shared reservations about overreliance on technology and the accuracy of AI-generated answers. These reservations resemble those previously noted in the literature. Therefore, while this study reveals that NotebookLM can support online learning processes, it emphasises the importance of a strategic framework for integration.

According to the findings obtained from the research, no significant difference was found between male and female students' academic achievement scores. This indicates that the students' readiness levels before the lesson are relatively like one another. Similarly, it is reported in the literature that significant differences do not always emerge according to the gender factor in technology-supported learning processes (Aesaert & van Braak, 2015). This finding is very important in the sense that the effects of NotebookLM are independent of the initial conditions. However, due to the limited sample size and cultural context differences, it is recommended that possible gender-based differences be examined with larger samples.

The fact that the development scores obtained after the NotebookLM application did not differ according to gender suggests that technology may have a gender-independent effect on learning outcomes in digitally supported learning processes. This result aligns with findings that artificial intelligence-based learning environments can be designed inclusively, mainly when artificial intelligence is utilized (Holmes et al., 2019). Moreover, studies emphasizing the importance of egalitarian approaches in educational technologies (Zawacki-Richter et al., 2019) suggest that the quality of pedagogical design is more effective than gender in determining learning outcomes. However, the fact that the findings are not statistically significant does not mean that the gender factor is completely ineffective; the results can be diversified with different course contexts or longer-term applications.

The significant increase in scores between the pretest and posttest reveals the enhancing effect of the NotebookLM application on academic achievement. This result aligns with the findings of similar studies, which suggest that AI-supported learning tools facilitate students' learning (Imran & Almusharraf, 2024; Peláez-Sánchez et al., 2024). It is seen that activities such as summarising, visualising and problem solving contribute positively to students' digital literacy skills. However, in terms of the generalisability of these findings, it is thought that it would be useful to examine the use of NotebookLM in different age groups, different course contents, and different durations. Furthermore, it is essential to investigate the impact of LLM-based tools on various variables, such as critical thinking and academic honesty, in future studies.

Student opinions indicate that NotebookLM has a positive impact on the learning process in various areas, including facilitating the understanding of course content, summarizing long texts, and thereby saving time, as well as contributing positively to student motivation. This result is in line with previous studies (Holmes et al., 2019; Zawacki-Richter et al., 2019), which state that AI-based learning tools can reduce the cognitive load, and students can learn the subject more effectively. Moreover, the increase in motivation expressed by the students can be explained by the personalised learning experiences offered by AI. Similarly, Peláez-Sánchez et al. (2024) state that LLMs motivate students by giving them learning autonomy. Despite all these results, there is a need for longer-term studies conducted in different courses to determine the permanence of students' positive attitudes towards the application.

Students generally reported that NotebookLM was user-friendly, easy to use, and that they did not encounter many technical problems during use. This result is consistent with the results of previous studies showing that perceived usefulness and ease of use are key concepts in the acceptance of new technologies (Aesaert & van Braak, 2015; Davis, 1989). Ease of use with a simple interface enables students to quickly adapt to the process and increases the usability of the technology for educational purposes. However, the contractions experienced in the system, especially when working with content with large file sizes, may negatively affect the user experience in the long term. For this reason, it would be an important orientation to improve the system technically and to continue continuous development activities by taking user feedback into account.

Students stated that NotebookLM's instant error recognition and personalised feedback reinforced learning. The results of this study agree with those of studies reporting that timely and constructive feedback provided to student agreeing process improves learning outcomes (Hattie & Timperley, 2007). Moreover, the fact that the application works in harmony with the teacher's notes is defined as an effect that facilitates teacher-student interaction, which contributes to the strengthening of the learning community. When examined in the literature, it is frequently stated that artificial intelligence-supported tools can play a supportive role in teaching processes, while it is emphasised that teachers cannot fulfil their social roles (Holmes et al., 2019; Zawacki-Richter et al., 2019). Within this scope, it is important to define NotebookLM as a tool to assist teachers in the teaching process.

This study not only presents descriptive results but also provides empirical insights on student experiences of LLM-based AI tools such as motivation, comprehension, or engagement. Therefore, the findings of the study not only contribute to the ever-expanding literature on AI-supported educational activities but also provide practical recommendations for the use of AI-based tools in higher education contexts.

Student feedback is primarily focused on the provision of Turkish language support and the inclusion of different interactive elements such as different visualisation possibilities and exam mode. The main reason for this finding may stem from the fact that students prefer outputs that include multimedia components instead of text-based outputs. Previous studies have stated that audiovisual materials have an important effect on permanent learning and that learning designs that include multimedia components contribute to the improvement of learning outcomes (Mayer, 2017). As a result, enriching NotebookLM with multimedia elements and interactive modules can appeal to a broader range of students with diverse learning styles. In addition, improving Turkish language support will contribute to a more effective and user-friendly experience for using the system throughout the country.

Another result considered important by the researchers is related to the students' tendency to recommend NotebookLM to their peers. This tendency can be considered as an indicator of students' adoption of this tool and their intention to reuse it. According to the Technology Acceptance Model, the intention to reuse a technology is an important indicator of its long-term use, that is, its sustainability (Davis, 1989). Therefore, students' intention to reuse NotebookLM emphasises the potential of NotebookLM as a learning tool that supports learning. However, the long-term use of the tool depends on continuous improvement of the vehicle in line with user feedback.

## 6 Conclusions

The quantitative results of the study revealed that NotebookLM had a positive effect on learning outcomes. The significant difference between pretest and posttest academic achievement scores suggests this. In addition, no significant difference was found between the academic achievement scores in terms of gender factor. This shows that NotebookLM has a consistent effect on all students regardless of gender. This result shows that artificial intelligence-based tools can enable an inclusive and egalitarian approach to online environments.

The qualitative findings are consistent with the quantitative results. Many students emphasised that the use of NotebookLM facilitated their learning process. Again, thanks to the summarising feature, it is stated that it contributes to both repeating the lessons easily and making the process more efficient. Additionally, the presence of these contributions enhances student motivation. Students can complete the course content effectively. They also stated that they had an improved learning experience thanks to the instant and personalised feedback they received. In line with this perspective, it can be said that the observed increase in students' academic achievement is related not only to cognitive but also to affective dimensions. These results, which are consistent with the existing literature, show that artificial intelligence-supported tools used in digital education environments can contribute to student development both in the cognitive sense and in the affective sense such as motivation.

In conclusion, when both qualitative and quantitative findings are evaluated together, it can be said that the use of the NotebookLM tool in the online digital literacy course has a multidimensional effect on learning outcomes. Students' personal experiences supported the significant increase observed in academic achievement. The facilitative effect on the learning process, user-friendly interface and positive outcomes related to motivation stand out as the main factors behind this success. Moreover, it can be stated that the lack of interaction, which is claimed as a limitation of online environments, can be overcome with the help of such tools. From this point of view, NotebookLM can be defined not only as a tool that increases success, but also as an educational technology that improves online learning experiences. However, student feedback, such as strengthening Turkish language support or adding interactive modules, reveal that the application should be continuously improved.

This study shows that the use of NotebookLM not only increases student academic achievement but also enriches online learning experiences. The study suggests some strategies for instructional design, institutional policies and future research to unlock this potential. The first suggestion for educators is that the capabilities of NotebookLM should be directly related to the learning objectives of the course. Again, guidance activities and feedback for students will contribute to increasing motivation and improve learning outcomes. In addition, in-class activities for the accuracy of the information shared by the AI will contribute to improving both students' digital literacy skills and critical thinking skills. From a pedagogical point of view, NotebookLM should be defined as a tool that supports the teaching process and facilitates learning.

One of the most prominent recommendations for decision makers is to strengthen the technical infrastructure. Again, strengthening the Turkish language support can be evaluated in this context. In-service training courses for teachers should be organised to effectively integrate the tool into educational environments. In addition, a responsible use approach should be adopted by considering data confidentiality, academic integrity and other ethical elements during the use of the tool. For researchers, longer-term impact studies, studies to be carried out with larger and different sample groups, and media comparative studies specific to different AI tools will contribute to a more accurate understanding of the effects of NotebookLM on educational activities. Therefore, the study provides empirical and pedagogical insights into the integration of AI-based learning tools into educational settings. It also goes beyond a descriptive account by emphasising their measurable impact on academic achievement and learning experience.

The findings of the study should be evaluated with the following limitations. This is considered important in terms of validity and generalisability. The study was conducted with a limited group of students in a single educational institution. This situation limits the external validity of the study. Secondly, the five-week intervention may not be sufficient to reveal the long-term and delayed effects of the study. In addition, the qualitative data of the study are based on student expressions. Although this data is considered valuable in terms of understanding student perceptions, further analysis may be necessary to comprehend cognitive processes.

Future studies should aim to investigate the long-term effects of using LLMs in educational settings. In addition, educator views and experiences towards AI-supported learning should also be revealed. Finally, ethical concerns such as bias, confidentiality, opaque decision-making processes and accountability can be considered. Through comparative and longitudinal research designs, more insight can be gained into the feasibility and sustainability of integrating AI into educational settings.

## 7 Competing Interests

The authors declare that there is no conflict of interest among them.

**Acknowledgements** We would like to express our gratitude to all students for their contribution to the study. Additionally, artificial intelligence tools such as ChatGPT and Grammarly were utilized primarily for language editing, enhancing textual flow, and translation purposes.

**Ethics** The study was conducted in accordance with the principles set out in the Helsinki Declaration. Each student was asked to complete an informed consent form. This manuscript has not been published anywhere or is under consideration for publication.

## Appendix A

### Sample Interview Questions

**General Experience and Use**

1. Could you briefly describe your experience when using NotebookLM for the first time?
2. For what purposes did you use the platform? Can you explain which feature you liked the most?
3. Were there any technical difficulties you encountered during the use of NotebookLM? Can you explain?

**Learning Process and Effectiveness**

1. Did NotebookLM facilitate your understanding of the topic?
2. What advantages did the platform provide you in your learning process?
3. Can you explain how the feedback provided by NotebookLM affects your learning?

**Communication and Interaction**

1. What was the impact of the content provided by your teachers during your interaction with NotebookLM?
2. How did the platform support your communication with your teacher?

**Suggestions for Development and Improvement**

1. What kind of improvements would you suggest for the more efficient use of NotebookLM?
2. What features would you like to see added to the platform to increase its contribution to your education process?

**Future Use and Recommendations**

1. Would you consider using NotebookLM again in the future? Why?
2. Would you recommend NotebookLM to your friends or other students? What are your reasons for recommending it?

# Appendix B

## Academic Achievement Test

1. **Which of the following can be considered both an advantage and a disadvantage of portable technologies?**
A. Use of multimedia
B. Flexibility to use anywhere
C. Rechargeable batteries
D. Issues arising from prolonged use
E. Personalization capability

2. **Which of the following cannot be said about lifelong learning?**
A. Lifelong learning is a process that covers all learning a person acquires throughout life.
B. In lifelong learning, control lies with the individual rather than the institution.
C. Lifelong learning emphasizes extending learning opportunities across all of life.
D. Lifelong learning is built on individuals being conscious, voluntary, and self-motivated in the learning process.
E. Lifelong learning means on-the-job training.

3. **Which of the following is not an advantage of users working with computers that have simpler hardware?**
A. Internet connectivity
B. Maintenance and repair cost
C. Hardware cost
D. Energy consumption
E. Noise

4. **Which of the following passwords is the least secure?**
A. ABC123
B. D@v1d5.
C. P@r1S.
D. D@v1d5
E.? Po9is*

5. **According to the network society approach, which of the following are the fundamental determinants transforming societies?**
A. Industrial development and manufacturing
B. Socio-cultural values
C. Computer technologies and information flows
D. Military technologies
E. Economy and production

6. **Which of the following is <u>not</u> one of the stages of technological development?**
A. Marketing
B. Innovation
C. Diffusion
D. Invention
E. Discovery

7. **Which of the following is not among the principles developed by the Computer Ethics Institute forming the basis of ethical rules for computer use?**
A. A computer must not be used to spread false information.
B. A computer must not be used in profit-making activities for people.
C. One must consider the social impacts of programs they write.
D. A computer must not be used to steal.
E. Other people's computer work must not be interfered with.

8. **Which of the following is not an aim of DoS attacks?**
A. Disabling a target by sending multiple packets from zombie machines
B. Penetrating the network and blocking authorized user access
C. Leaving a backdoor on the target to ease future intrusion
D. Overloading a server with excessive requests to render it inoperable
E. Capturing sensitive information

9. **Which language is used to create web pages on the Internet?**
A. http
B. URL
C. www
D. ftp
E. html

10. **Considering humanity's technological evolution, the earliest tools primarily aimed to provide which of the following?**
A. Building shelters
B. Seeking food and protection from predators
C. Practicing animal husbandry
D. Transitioning to agricultural production
E. Protection from other humans

11. **Which of the following is <u>not</u> among the economic goals of Industry 4.0?**
A. Occupational safety
B. High standard of living
C. Diminishing returns
D. Economic growth
E. Employment and social stability

12. **Which of the following is <u>not</u> among the digital rights defined for children and youth?**
A. Right to Digital Safety and Support
B. Right to Sell Personal Data
C. Right to Know

D. Right to Erasure
E. Right to Digital Literacy Skills

13. **As a wiki, what is Wikipedia's disadvantage in terms of reliability?**
A. Everyone cannot view its content
B. Only teachers can write articles
C. It exists only as printed material
D. There is no mechanism to verify information reliability
E. It is difficult to update

14. **Which feature of consumer law places it among the "branches of law that protect the weaker party"?**
A. Increasing social responsibility awareness
B. Adapting to technological innovations
C. Supporting environmental sustainability
D. Ensuring economic balance
E. The need to protect consumers

15. **What is Artificial Intelligence (AI)?**
A. A computer programming language
B. An internet browser
C. A hardware unit
D. The science of developing machines with human-like thinking and learning abilities
E. A tool used only for data analysis

16. **Deep learning operates using which structure?**
A. Random forests
B. A natural language processing model
C. Neural networks
D. Support vector machines
E. Decision trees

17. **In general, entrepreneurship cannot be associated with which of the following?**
A. Creating new opportunities
B. Organizing the inputs of production
C. Taking risks
D. Creating crises
E. Exploiting emerging opportunities

18. **Thanks to mobile applications, businesses gain which advantage?**
A. Slow and limited access
B. Need for a physical storefront
C. Ability to deliver continuous and high-quality service
D. High costs and headcount
E. Traditional advertising methods

19. **Which of the following is <u>not</u> one of the factors of the entrepreneurial climate that varies from country to country?**
A. Structure of state governance
B. Development of labor and financial markets
C. Gender differences
D. Levels of technology use
E. Differences in social and cultural infrastructure

20. **Among the advantages of using the Internet for educational purposes, which is emphasized?**
A. User privacy issues
B. Limited access opportunities
C. High costs
D. Risk of encountering inappropriate content
E. Removal of time and place constraints

21. **Why is the security of IoT devices important?**
A. To make devices run faster
B. To extend battery life
C. To connect more devices to the internet
D. To prevent data breaches and cyberattacks
E. To commit personal data violations

22. **According to the European Commission's "Digital Competence Framework for Citizens," which of the following is not among the five core competences?**
A. Accessing digital data and information
B. Financial management
C. Safety
D. Communication and collaboration
E. Digital content creation

23. **Which query should be used in a search engine to find web pages where the phrase "Açıköğretim Fakültesi" appears <u>exactly</u> as written?**
A. Açıköğretim+Fakültesi
B. Açıköğretim AND Fakültesi
C. Açıköğretim*Fakültesi
D. Açıköğretim Fakültesi
E. "Açıköğretim Fakültesi"

24. **Which of the following describes a MOOC?**
A. An example of Web 2.0 tools
B. Massive Open Online Courses
C. An instructional method used only in physical classrooms
D. Another name for traditional classroom teaching
E. A learning tool used only on mobile devices

25. **What is Augmented Reality (AR)?**
A. A technology used to store medical records
B. A technology created by overlaying computer-generated images, sounds, or other data onto the real-world environment
C. An offline security system
D. A term used only in the gaming industry
E. A technology where users feel themselves in a completely virtual environment

## References

Aesaert, K., & van Braak, J. (2015). Gender and socioeconomic-related differences in performance-based ICT competences. *Computers & Education, 84*, 8–25. https://doi.org/10.1016/j.compedu.2014.12.017

Alanoğlu, M., & Karabatak, S. (2020). Artificial intelligence in education. In F. Y. Güçlü & M. N. Kaymak (Eds.), *Educational research* (pp. 175–185). Eyuder Publications.

Alisoy, H. (2025). Can NotebookLM support English language learners? A theoretical perspective on AI tools in education. *Porta Universorum, 1*(6), 25–55. https://doi.org/10.69760/portuni.0106003.

Berg, B. L., & Lune, H. (2012). *Qualitative research methods for the social sciences* (8th ed.). Pearson.

Bui, H. P., Bui, H. P. H., & Dinh, P. D. (2023). Vietnamese students' use of smartphone apps in English learning. *LEARN Journal: Language Education and Acquisition Research Network, 16*(1), 28–46. https://so04.tci-thaijo.org/index.php/LEARN/article/view/263430

Bui, H. P., & Vo, L.-H. (2024). Interviews: Considerations and techniques. In H. P. Bui (Ed.), *Considerations and techniques in applied lingusitics and language education research* (pp. 136–145). IGI Global.

Büyüköztürk, Ş., Kılıç-Çakmak, E., Akgün, Ö., Karadeniz, Ş., & Demirel, F. (2018). *Scientific research methods*. Pegem Academy. https://doi.org/10.14527/9789944919289.

Chisega-Negrilă, A. M. (2024). Teaching and learning in an AI-powered world. *Bulletin of "Carol I" National Defence University, 13*(3), 105–116. https://doi.org/10.53477/2284-9378-24-33

Chukwuere, J. (2024). Rapid review of the COVID-19 pandemic's impact on the digitalization of higher education. *Research on Education and Media, 16*(1), 1–9. https://doi.org/10.2478/rem-2024-0002

Cohen, E., & Nycz, M. (2006). Learning objects and e-learning: An informing science perspective. *Interdisciplinary Journal of E-Learning and Learning Objects, 2*(1), 23–34. https://doi.org/10.28945/399.

Cohen, J. (1988). *Statistical power analysis for the behavioral sciences* (2nd ed.). Lawrence Erlbaum Associates.

Creswell, J. W. (2021). *A concise introduction to mixed methods research* (2nd ed.). SAGE Publications.

Davis, F. D. (1989). Perceived usefulness, perceived ease of use, and user acceptance of information technology. *MIS Quarterly, 13*(3), 319–340. https://doi.org/10.2307/249008

Filippi, S., & Motyl, B. (2024). Large Language Models (LLMs) in engineering education: A systematic review and suggestions for practical adoption. *Information, 15*(6), 345. https://doi.org/10.3390/info15060345

Goyal, S. (2012). E-Learning: Future of education. *Journal of Education and Learning, 6*(4), 239–242. https://doi.org/10.11591/edulearn.v6i4.168.

Hattie, J., & Timperley, H. (2007). The power of feedback. *Review of Educational Research, 77*(1), 81–112. https://doi.org/10.3102/003465430298487

Holmes, W., Bialik, M., & Fadel, C. (2019). *Artificial intelligence in education: Promises and implications for teaching and learning*. Center for Curriculum Redesign.

Hwang, G.-J., Tu, Y.-F., & Tang, K.-Y. (2022). AI in online-learning research: Visualizing and interpreting the journal publications from 1997 to 2019. *The International Review of Research in Open and Distributed Learning, 23*(1), 104–130. https://doi.org/10.19173/irrodl.v23i1.6319.

Imran, M., & Almusharraf, N. (2024). Google Gemini as a next generation AI educational tool: A review of emerging educational technology. *Smart Learning Environments, 11*(1), 22. https://doi.org/10.1186/s40561-024-00310-z

Ivanov, S. (2023). The dark side of artificial intelligence in higher education. *The Service Industries Journal, 43*(15–16), 1055–1082. https://doi.org/10.1080/02642069.2023.2258799

Jethro, O. O., Grace, A. M., & Thomas, A. K. (2012). E-learning and its effects on teaching and learning in a global age. *International Journal of Academic Research in Business and Social Sciences, 2*(1), 244–252.

Khoa, B. T., Hung, B. P., & Hejsalem-Brahmi, M. (2023). Qualitative research in social sciences: data collection, data analysis and report writing. *International Journal of Public Sector Performance Management, 12*(1,2), 187–209. https://doi.org/10.1504/IJPSPM.2023.132247.

Klisowska, I., Seń, M., & Grabowska, B. (2020). Advantages and disadvantages of distance learning. *E-methodology, 7*(7), 27–32. https://doi.org/10.15503/emet2020.27.32

Kuimova, M., Kiyanitsyna, A., & Truntyagin, A. (2016). E-learning as a means to improve the quality of higher education. *SHS Web of Conferences, 28*, 01129. https://doi.org/10.1051/shsconf/20162801129

Luckin, R., Holmes, W., Griffiths, M., & Forcier, L. B. (2016). *Intelligence unleashed: An argument for AI in Education.* Pearson.

Mayer, R. E. (2017). Using multimedia for e-learning. *JCAL, 33*(5), 403–423. https://doi.org/10.1111/jcal.12197.

McCarthy, J. (2007). *What is artificial intelligence?* Stanford University, Formal Reasoning Group. https://cse.unl.edu/~choueiry/S09-476-876/Documents/whatisai.pdf

Miranda, S. (2025). Artificial Intelligence from Google environment for effective learning assessment. *Information, 16*(6), 462. https://doi.org/10.3390/info16060462

Ojha, P., & Kadel, L. B. (2024). A comparative study of different machine learning based AI tools. *International Journal of Science and Research, 13*(3), 1092–1094. https://doi.org/10.21275/SR24314225540.

Özkan, Y. (2019). *Artificial Intelligence Workshop in Distance Learning.* [Final report]. TASAM. https://tasam.org/tr-TR/Icerik/51444/uzaktan_ogretimde_yapay_zekâ_egitimde_yapay_zekâ_calistayi_-_5_sonuc_raporu

Peláez-Sánchez, I. C., Velarde-Camaqui, D., & Glasserman-Morales, L. D. (2024). The impact of large language models on higher education: Exploring the connection between AI and Education 4.0. *Frontiers in Education, 9*, 1392091. https://doi.org/10.3389/feduc.2024.1392091.

Preniqi, N., Gërvalla, M., & Sylaj, K. (2020). Digital transformation in e-learning education. *UBT International Conference, 1.* https://knowledgecenter.ubt-uni.net/conference/2020/iss/1

Rabadanova, R., Seminskaya, E., & Filatov, A. (2024). The negative aspects of using artificial intelligence for students in solving individual educational tasks. *Applied Psychology and Pedagogy, 9*(3), 115–124. https://doi.org/10.12737/2500-0543-2024-9-3-115-124.

Radović-Marković, M. (2010). Advantages and disadvantages of e-learning in comparison to traditional forms of learning. *Annals of the University of Petroşani, Economics, 10*(2), 289–298.

Şenocak, D. (2020). Artificial intelligence in open and distance learning environments: Opportunities and concerns. *Journal of Open Education Applications and Research, 6*(3), 56–78.

Shahzad, T., Mazhar, T., Tariq, M. U., Ahmad, W., Ouahada, K., & Hamam, H. (2025). A comprehensive review of large language models: Issues and solutions in learning environments. *Discover Sustainability, 6*(1), 27. https://doi.org/10.1007/s43621-025-00815-8

Shet, J. P. (2020). Adaptive and blended learning the panacea for the challenges of e-learning. *JETIR, 7*(5), 707–720. https://www.jetir.org/papers/JETIR2005241.pdf

Stecuła, K., & Wolniak, R. (2022). Advantages and disadvantages of e-learning innovations during COVID-19 pandemic in higher education in Poland. *Journal of Open Innovation: Technology, Market, and Complexity, 8*(3), 159. https://doi.org/10.3390/joitmc8030159

Swacha, J., & Gracel, M. (2025). Retrieval-Augmented Generation (RAG) chatbots for educational purposes: A review of 47 papers. *Applied Sciences, 15*(8), 4234. https://doi.org/10.3390/app15084234

Ülkü, A. (2023). Artificial intelligence-based large language models and integrity of exams and assignments in higher education: The case of tourism courses. *Tourism & Management Studies 19*(1), 21–34. https://doi.org/10.18089/tms.2023.190402.

Vincent-Lancrin, S., & Van der Vlies, R. (2020). Trustworthy artificial intelligence (AI) in education: Promises and challenges. *OECD Education Working Papers, 218*, 1–17. https://doi.org/10.1787/a6c90fa9-en.

Wollny, S., Schneider, J., Di Mitri, D., Weidlich, J., Rittberger, M., & Drachsler, H. (2021). Are we there yet? A systematic literature review on chatbots in education. *Frontiers in Artificial Intelligence, 4*, Article 654924. https://doi.org/10.3389/frai.2021.654924

Yusuf, N., & Al-Banawi, N. (2013). The impact of changing technology: The case of e-learning. *Contemporary Issues in Education Research, 6*(2), 173–180. https://doi.org/10.19030/CIER.V6I2.7726.

Zawacki-Richter, O., Marín, V. I., Bond, M., & Gouverneur, F. (2019). Systematic review of research on artificial intelligence applications in higher education—Where are the educators? *International Journal of Educational Technology in Higher Education, 16*(1), 39. https://doi.org/10.1186/s41239-019-0171-0

**Bünyami Kayali** holds a degree in Primary School Mathematics Teaching. He has a master's degree and a doctorate in Computer Education and Instructional Technologies. He currently works at Bayburt University. His research areas include Distance Education, Emergency Distance Education Instructional Technologies, and Artificial Intelligence.

**Mehmet Yavuz** holds a degree in Computer Education. He has a master's degree and a doctorate in the Department of Computer Education and Instructional Technologies. He currently serves as an associate professor at Bingöl University. His research interests include digital transformation, distance education, instructional design, teacher education, instructional technologies, and artificial intelligence.

# Impacts of Chatbot Training on EFL Teachers' Confidence and Acceptance of AI Technologies

**Nguyen Truong Sa, Nguyen Thi Diem Thi, Ho Thi Hien, and Hoang Thanh Dung**

**Abstract** Technological revolutions have huge impacts on various disciplines of life, including language education. Therefore, the employment of new technologies like artificial intelligence (AI) has been the focus of both research and practice. However, as basic practitioners, most language teachers' acceptance and use of EdTech seems fundamental and basic. Hypothesizing that training would change the teachers' confidence and acceptance levels in using AI chatbots, this study examined such a relationship would change among a group of 27 Vietnamese teachers of English before and after being trained and coached on developing these tools. We adopted a mix-methods research design and adapted the UTAUT2 framework. We used a set of surveys to collect quantitative data to investigate the extent of the change. Focused group interviews with the participants were organized during a 3-week training period to gain insights into the change. Employing T-test, Cohen's *d* effect size and content analysis, the study found that the trained teachers' confidence varied. Also, their acceptance of the chatbot significantly increased after the traning. The findings suggest that adequate training could greatly benefit foreign language teachers to accept EdTech.

**Keywords** AI chatbot · Educational technologies · Teacher's acceptance · Teacher's confidence · UTAUT2

## 1 Introduction

Teaching and learning practices have undergone significant transformation as a response to the rapid evolution of AI in recent years (Chen et al., 2020; Huang et al., 2021), driving an urge for language teachers to prepare themselves with profound professional digital competence to foster teaching and learning. However, a large number of teachers tend to play a rather passive role in technology employment.

N. T. Sa (✉) · N. T. D. Thi · H. T. Hien · H. T. Dung
Industrial University of Ho Chi Minh City, Ho Chi Minh City, Vietnam
e-mail: nguyentruongsa@iuh.edu.vn

 
H. P. Bui and L. P. H. Huong (eds.), *Artificial Intelligence in Education and Academic Research*, Studies in Computational Intelligence 1269,
https://doi.org/10.1007/978-3-032-18392-7_4

They primarily depend on pre-installed functions and usually have to make modifications to adapt apps or websites to their intended purposes. According to UNESCO (2023), the teachers' role is very limited in making technology-related decisions, with 45% of teachers in 94 countries reporting being required to work with new technology without adequate consultation. This process can be complicated and tiring for those with limited technological expertise, such as teachers. Therefore, it is essential to prepare teachers to be more pro-active in technology, especially AI. Yet with a purely linguistics background, teachers' confidence in receiving AI training and the possibility of applying it in practice is still an under-researched topic. To close these theoretical and practical gaps, this study investigated how teachers' confidence and acceptance might change when they are coached to use a technology beyond their trained expertise, specifically in this study—a chatbot. Chatbots were chosen as the subject of this study because developing an effective chatbot is a challenging task requiring extensive technological knowledge (Yan et al., 2016), thereby allowing the researchers to observe the real impact of training on non-specialists such as teachers.

To achieve the aforementioned aim, the study would try to answer two research questions:

1. How will training affect teachers' confidence when using AI chatbots?
2. How will teachers' acceptance of AI chatbots change after training?

## 2 Literature Review and Theoretical Framework

### *2.1 AI Chatbots in Education*

AI chatbots, computer programs that replicate human speech through natural language processing—are increasingly being utilized in educational settings to promote learning (Labadze et al., 2023). In the realm of language learning, as asynchronous conversation partners, chatbots can provide real-time, responsive engagement for customized practice without the pressure of judgment (Panagiotidis, 2024). In their systematic review, Du and Daniel (2024) acknowledged the value AI chatbots could offer in advancing speaking skills and alleviating foreign language related anxiety; however, further research in varied educational contexts is necessary. In addition, a thorough meta-synthesis documenting the use of chatbots in the instruction of communication skills and integration of other related studies from 2010 to 2024 illustrates how learners speaking skills, vocabulary and overall motivational toward the language learning increased. The reason for such improvement is due to the genuine conversations chatbots simulate, the instant personalized feedback they offer, and the near real-life dialogues they maintain (Ding & Yusof, 2025; Kim & Su, 2024; Koç & Savaş, 2025; Wang et al., 2024; Wollny et al., 2021).

## *2.2 Predicting User Acceptance of AI Chatbots in Language Teaching*

The motivation to utilize AI chatbots in language education stems from the predictive capabilities of Venkatesh et al. (2012) Unified Theory of Acceptance and Use of Technology 2 (UTAUT2) and its subsequent additions. The theory contains psychological, societal, and organizational components (Venkatesh et al., 2012, 159), namely performance expectancy (PE), effort expectancy (EE), social influence (SI), and facilitating conditions (FC), that predict one's intention to adopt a certain technology. Self-efficacy (SE) has been added in this study's educational context, considering the user's psychological disposition and emotional readiness to adopt the technology (Kent & Giles, 2017).

In this study, to analyze language teaching educators' acceptance of AI chatbots, an adapted UTAUT2-based framework (Venkatesh et al., 2012) is applied. This framework has been customized to fit the study's pedagogical and technological context, as illuminated in Fig. 1. The model comprises eight constructs from which the researcher has designed the framework: "PE, EE, SI, FC, SE, Hedonic Motivation (HM), Habit (HA), and Perceived Risk (PR)." All of these constructs are theorized to impact the variable, Behavioral Intention (BI).

This model preserves the UTAUT2 pillars: PE, EE, SI, and FC. These constructs are well established as predictors of technology adoption across diverse contexts (Venkatesh et al., 2012, p. 159). In addition, two constructs from UTAUT2 adapted for the consumer setting—HM and HA were added. HM is "the fun or pleasure derived from using technology" (Venkatesh et al., 2012, p. 8), which may reflect the enjoyment or satisfaction that users derive from engaging with the chatbot. HA, on the other hand, describes how the access and use of the chatbot becomes habitual. HA is defined as "the extent to which people tend to perform behaviors automatically

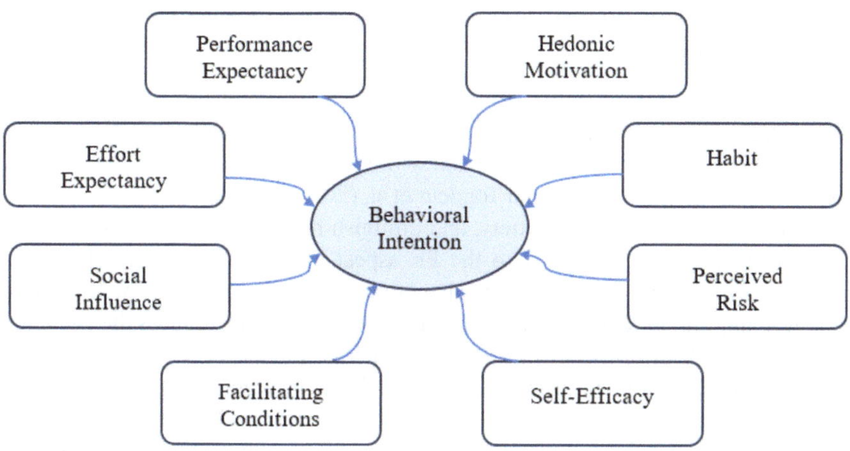


**Fig. 1** Adapted theoretical framework for AI chatbot acceptance

because of learning" (Limayem et al., 2007, p. 709) and can act as both a direct and indirect driver of technology use (Venkatesh et al., 2012). These modifications are in line with prior technology acceptance model modifications in educational settings (Amer jid Almahri et al., 2024). SE, one of the missing constructs in the UTAUT2 framework but covered in earlier educational research, for example, has also been added to address teachers' perceived competencies with the chatbot (Moghavvemi, 2015). BI, the dependent construct, describes the teachers' inclination to use the chatbot in their lessons. As in other UTAUT-based studies, BI is viewed as the most proximate predictor of actual technology use (Venkatesh et al., 2012).

Although the concept of confidence may not appear relevant to the UTAUT2 model, it was argued to be aligned closely with SE (Moghavvemi, 2015). The original UTAUT model relegated self-efficacy to an indirect construct, but SE is considered an important antecedent influencing core UTAUT determinants (Moghavvemi, 2015; Tang et al., 2024). Research based on extended UTAUT models shows that SE positively influences EE and PE and is a predictor of FC (Moghavvemi, 2015; Tang et al., 2024). In the context of technology adoption, technology users' confidence reflects how strongly users believe they can complete the necessary tasks and positively transforms a user's perception concerning the ease of use (EE/PEOU) and the usefulness (PE/PU) of the technology (Tang et al., 2024). Bandura (1994, 1997) suggested that confidence leads to higher self-efficacy, which positively impacts the user's intention (BI) by alleviating self-efficacy anxiety, increasing persistence, and overcoming the obstacles to technology, and giving the user a higher sense of control over the technology. Therefore, SE can be treated as a motivational construct concerning technology acceptance, and robust SE beliefs are needed for the effective use of technology. This construct is crucial in explaining the psychological aspects of the BI changes when adopting newer technologies like AI chatbots.

## 2.3 The Role of Technological Training for Language Teachers

According to recent studies, appropriately structured, experiential training significantly increases teachers' confidence and willingness to use educational technology. Buchner and Hofmann (2022) and Tondeur et al. (2025) proposed that training workshops increase SE, decrease anxiety, and diminish barrier perception. These confidence gains closely correspond to the PE aspect of UTAUT2, and teachers who believe they can use technology are also convinced that technology will improve their teaching. Importantly, scaffolded and immersive training reduces the challenge of the task which, in turn, improves EE. Nevertheless, Tondeur et al.'s (2025) research indicated that even trained teachers still do not have the readiness to employ digital technology in their teaching. In studies such as Buchner and Hofmann (2022), Hsu and Lin (2020) and Tondeur et al. (2025), effects of technical training workshops focusing on specific apps or devices on changing classroom practices were very minimal or unidentified. Therefore, there is an urgent need for intensive professional development (Amemasor et al., 2025; Hirsch et al., 2018). To achieve learning aims, understand the teacher's context, and promote active collaborative reflection and critical self-reflection, technological teacher training programs should also focus more on outcomes of learning rather than outcomes on tools.

Despite the abundant literature highlighting the opportunities afforded by EdTech and AI chatbots in the language education sector, several gaps remain. First, the literature over-focused on the adoption of AI chatbots rather than on teachers' use and attitudes. This is even more pronounced in the understudied context of Vietnam, where teachers' readiness and confidence to use AI chatbots are highly debated. The existing literature is largely centered on students and the outcomes of learning, thus ignoring the adoption processes of teachers in professional development, teacher attitudes, and practices. Finally, there is a gap for researching impacts of these training models on teacher confidence in and acceptance of the use of conversational AI tools. The lack of studies in the field of AI technology that build or adapt the UTAUT2 model to include SE and PR also remains a significant gap in the literature.

Moreover, there still has been little focus on how short, targeted coaching sessions influence a teacher's technological SE as well as their willingness to integrate AI tools in teaching. This research addresses these gaps by investigating how AI tools, particularly chatbots, can be integrated in a confident and goal-directed manner in EFL teaching during the systematic training that highlights the balance of technological and pedagogical integration.

# 3 Research Design

## *3.1 Research Design*

This study employed a mixed methods research design (Riazi, 2017) to examine the immediate effects of a chatbot training program on EFL teachers' acceptance of chatbots for teaching. A single-group pre-survey and post-survey experiment was conducted in a project where all the participants received the same chatbot training in a lab room followed by 3 weeks to build a chatbot for their teaching needs. The selection of this research design allows the researchers to observe and measure the participants' levels of chatbot acceptance after engaging in the training over time. It is also practical and useful in educational settings where including control groups for professional development projects may not be feasible.

## 3.2 Participants and Context

We adopted a purposive sampling method (Riazi, 2017; Tin & Bui, 2024) to select the target participants for the study. The participants were required to be current EFL teachers working either in public and private schools or language centers. They were asked to be available to attend a five-hour training session and complete the surveys. Invitations and consent were sent to EFL in-service teachers from different educational institutions in Southern Vietnam. Twenty-seven teachers met the requirements and confirmed their availability to join the chatbot training. The participants were between 24 and 35 years old. Their teaching experience ranges from 2 to 13 years. Their academic qualifications vary from BA and MA. None of them have participated in a chatbot training session before.

## 3.3 Training Session

A 5 h training session was coached by a group of trainers whose expertise is in chatbots developing services. They utilised the "ReallyBot" platform to allow users to create conversations and operate virtual agents who can communicate with users via text-based prompts integrated with online sources such as videos, links, and documents. ReallyBots-powered chatbots can help users automobilize the Q&A sessions and tutor their own learning. More important, "RealyBot" is typically known for its user-friendly interface without any required coding skills. The aim of the training was to provide hands-on guidance for all participants to build their own chatbots for teaching and learning purposes. Through this training session, participants understand safe, effective, pedagogical principles that a large-language-model (LLM) chatbot works.

The trainers followed a rigorous step-by step procedure during the training session to make sure that it is explicit and easy for all participants. During the training, participants were asked to think of some ideas how their chatbot can be created to assist their teaching or personalize their professional tasks- and develop the Storyboard to build the conversational flow. These training techniques and procedures were proposed by Nguyen et al. (2024). Finally, the participants were asked to plan the structure and purposes of a chatbot in their actual teaching settings (i.e. a chatbot to help learners self-study vocabulary, to explain a grammatical point). The trainers provided timely support for the initial chatbot creation of the participants regarding the arising problems in the trial running. After the training, participants have three weeks to complete their own chatbot. Some sample chatbot that participants created are shown in Figs. 2 and 3. To support the internal validity, all the researchers kept the attendance logs, a trainer checklist, and a collection of teacher-produced artifacts (e.g., prompt templates and task plans).

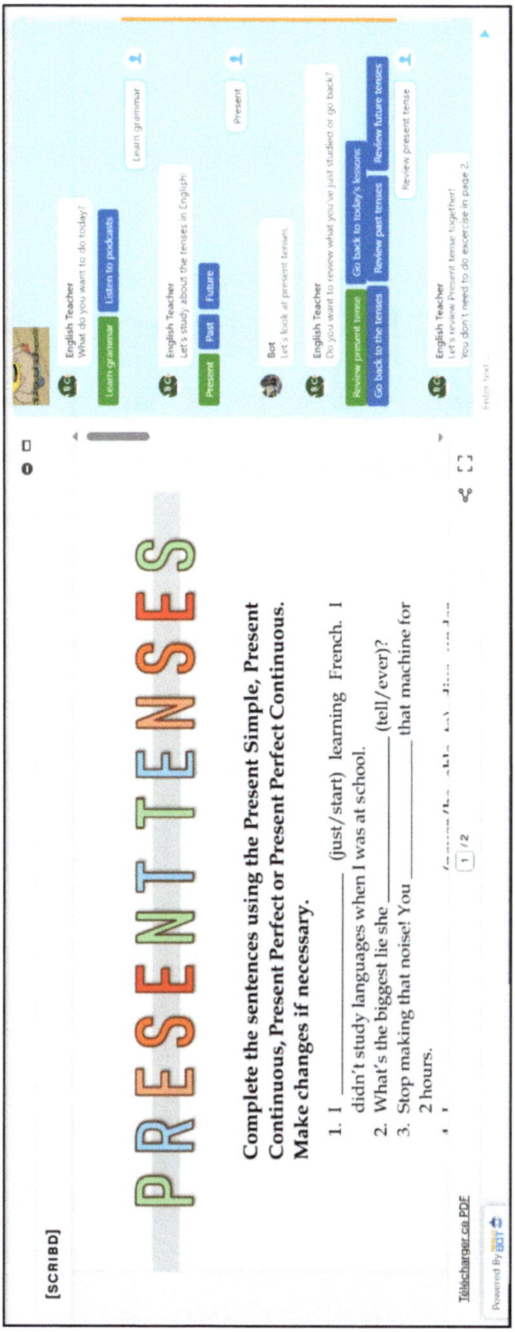


Fig. 2 Chatbot 1 created by a participant

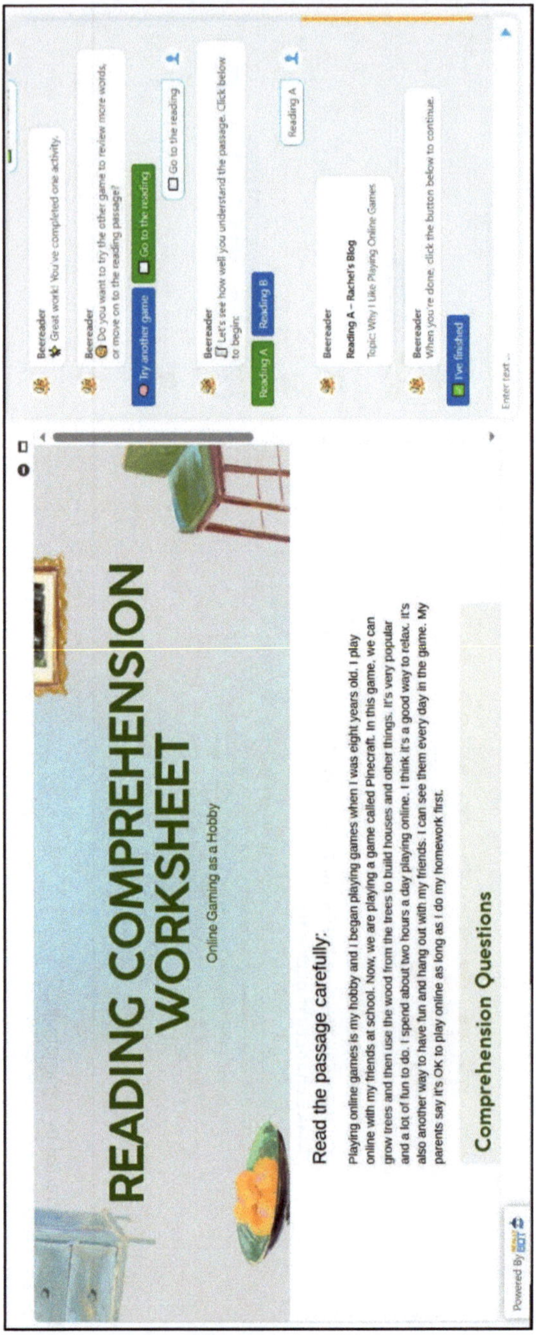


**Fig. 3** Chatbot 2 created by a participant

**Table 1** Internal consistency reliability (Cronbach's alpha) of constructs in the pre- and post-survey (N = 27)

| | PE | EE | SI | FC | HM | HA | PR | SE |
|---|---|---|---|---|---|---|---|---|
| Pre-survey | 0.940 | 0.891 | 0.847 | 0.854 | 0.922 | 0.915 | 0.656 | 0.825 |
| Post-survey | 0.910 | 0.890 | 0.987 | 0.906 | 0.921 | 0.925 | 0.821 | 0.873 |

### *3.4 Instruments*

Data was collected from pre-survey and post-survey to measure confidence and acceptance levels of chatbot creation and implementation. We employed Chau et al.'s (2024) proposal that a questionnaire could be used as a measure in social research. The adapted inventory of UTAUT2 (Venkatesh et al., 2012) requires participants to reflect on their experience and confidence levels with the trained chatbot. The main parts of the inventory included 25 Likert-scale items focusing on different constructs to measure participants' reflection of acceptance in chatbot implementation, namely PE with 4 items; EE with 3 items; SI with 3 items; FC with 3 items; HM with 3 items; HA with 3 items; PR with 3 items; SE with 3 items. Table 1 shows the Cronbach's alpha values in both of the pre-survey and post-survey. Acceptable to excellent reliability was observed in all of the constructs, with alpha coefficients ranging from 0.656 to 0.940 in the pre-survey and from 0.821 to 0.987 in the post-survey.

Furthermore, open-ended questions were employed to investigate more thoughtfully on chatbot creation experience such as satisfactory features, challenges, improvement and confidence.

Under Dörnyei and Taguchi's (2009) guidelines, a rigorous procedures were strictly followed to manage the constructing, piloting and validating the survey. The procedures include: drafting the English version based on UTAUT2 (Venkatesh et al., 2012) (Researcher 1), translating, polishing wording and validating content (Researcher 2 and 3). The final draft was piloted by 2 teachers who would not engage in the training for further feedback and revisions. Finally, the official surveys were administered to participants in Google Forms.

### *3.5 Data Collection and Analysis*

The researchers recruited the participants by sending a brief introduction and consent to the teachers. There were 40 registrations, but only 27 of them were able to show up on the training day. Second, participants were invited to complete the online pre-survey just before the training. Third, a training session was hosted in a Lab room. Next, the participants had three weeks to create their own chatbot, then the artifacts were collected. Finally, post-survey was administered as soon as the participants completed.

For research question 1, a paired-sample T-test was analysed to compare difference in means between the 2 surveys. The assumption is a higher score in the post-survey indicates an increase of the confidence levels in using the chatbot. In addition, content analysis was also used for data from the open-ended questions were analysed under the light of the pre-determined constructs of the adapted UTAUT2. To answer research question 2, Pearson correlation test between BI and other constructs was run to investigate the associations between participants' BI in their future teaching. Cohen's d analysis for paired samples was also calculated to quantify the effect size of possible changes. All analyses were conducted in SPSS (version 26).

# 4 Results and Discussion

## *4.1 How Training Affected Confidence Among Participants in Using the Chatbot*

As shown in Tables 2 and 3, participants reported a high level of satisfaction with the training, 23/27 of them reported a considerable change in confidence, the remaining 4 trainees felt little confidence, needing more training and instruction. When asked to evaluate the general benefits of training, most of them admitted that the training was quite effective in terms of equipping them with a basic knowledge of how a chatbot operates and providing hand-on skills on developing a simple chatbot. On the continuum Likert scale from 1 (not effective at all) to 7 (highly effective), all of the participants evaluated the training effectiveness from level 4; more than 50% of them (15 trainees) even rated the training at 6 and 7 levels. Effectiveness confirmed among the trainees was also strongly supported by their confidence in the possibility of explaining and guiding other people to develop a similar chatbot to the one they had built. Table 3 shows a large percentage of participants chose from level 5 on the confident continuum, which signifies that they highly believed in their potential output of re-training such guided content and processes.

However, based on the paired samples T-test (Table 4), when comparing 27 trainees' responses against the survey in post-survey and pre-survey in more detail,

**Table 2** Evaluation on the effectiveness of the training

| | | Frequency rate | Percent (%) | Valid percent (%) | Cumulative percent (%) |
|---|---|---|---|---|---|
| Valid | 4 | 4 | 14.8 | 14.8 | 14.8 |
| | 5 | 8 | 29.6 | 29.6 | 44.4 |
| | 6 | 14 | 51.9 | 51.9 | 96.3 |
| | 7 | 1 | 3.7 | 3.7 | 100.0 |
| | Total | 27 | 100.0 | 100.0 | |

**Table 3** Confidence in training other people to use

| | | Frequency rate | Percent (%) | Valid percent (%) | Cumulative percent (%) |
|---|---|---|---|---|---|
| Valid | 3 | 1 | 3.7 | 3.7 | 3.7 |
| | 4 | 2 | 7.4 | 7.4 | 11.1 |
| | 5 | 8 | 29.6 | 29.6 | 40.7 |
| | 6 | 10 | 37.0 | 37.0 | 77.8 |
| | 7 | 6 | 22.2 | 22.2 | 100.0 |
| | Total | 27 | 100.0 | 100.0 | |

the results are quite mixed with both negative and positive changes in the evaluation. Moreover, statistical significance was not observed ($p > 0.05$) in 13 out of 27 participants (student coded as number 1, 3, 7, 9, 11, 14, 18, 20, 21, 22, 23, 25 and 26). Nearly 41% of the participants (11/27) showed negative changes; the changes varied from—1.48276) to slightly negative (0.03448) after comparing the test mean scores. For the larger percentage (59%) of participants showing positive changes in mean scores, the changes ranged from 0.2069 to 1.58621 and most of the changes were higher than 0.500.

Analysis of the interview data with participants revealed more insights for the changes in confidence. According to participant N. 27, who noted a positive evaluation of the training effectiveness and a higher confidence after training, but had a negative result when pairing the T-test, "I finally can develop my storyboard and run the chatbot after 2 weeks. I know new skills from the training, I'm satisfied with the training, but I'm much slower and need more instructions than other people… I'm not very happy with my chatbot, it's still very basic and boring". For another trainee whose T-test score dropped, the reason was "I understand the guided process and the expectation on me, but in practice people will interact with a chatbot with more complicated queries and needs, I'm not sure I can manage to develop such a more interactive chatbot" (N. 16). Meanwhile, for trainees having positive changes in mean scores, despite knowing that the completed chatbot may still be uncompleted and needs improvements, their evaluations focused more on the benefits of new knowledge and skills they learned about chatbot. Students with positive changes in general mean scores believed that if they had been given more time and intensive training, they would have developed a more qualified chatbot.

> I was quite nervous at first, as I was thinking about working with coding and algorithms in which I have no ideas about. But then developing a chatbot was much simpler than I expected; … and with this platform, I do not need any background in coding (N. 8).

Recent literature posits that professional development can foster language educators' technological expertise (Alwaqdani, 2024; Amemasor et al., 2025; Ifenthaler et al., 2024; Liu, 2025; Pokrivčáková, 2019). Historically, educators' professional development has been aimed at helping teachers acquire the necessary technical and pedagogical competencies for the integration of educational technologies that

**Table 4** Paired samples test *comparing results*

| | | Post-survey and pre-survey paired differences | | | | | t | df | Sig. (2-tailed) |
|---|---|---|---|---|---|---|---|---|---|
| | | Mean | Std. deviation | Std. error mean | 95% confidence interval of the difference | | | | |
| | | | | | Lower | Upper | | | |
| Pair 1 | A2_A–A1_A | −0.06897 | 1.25160 | 0.23242 | −0.54505 | 0.40712 | −0.297 | 28 | 0.769 |
| Pair 2 | B2_A–B1_A | 0.41379 | 1.05279 | 0.19550 | 0.01333 | 0.81425 | 2.117 | 28 | 0.043 |
| Pair 3 | A2–A1 | 0.10345 | 1.54330 | 0.28658 | −0.48359 | 0.69049 | 0.361 | 28 | 0.721 |
| Pair 4 | B2–B1 | −1.48276 | 1.27113 | 0.23604 | −1.96627 | −0.99925 | −6.282 | 28 | 0.000 |
| Pair 5 | C2–C1 | 0.75862 | 0.87240 | 0.16200 | 0.42678 | 1.09046 | 4.683 | 28 | 0.000 |
| Pair 6 | D2–D1 | −0.93103 | 1.06674 | 0.19809 | −1.33680 | −0.52527 | −4.700 | 28 | 0.000 |
| Pair 7 | E2–E1 | 0.20690 | 1.31961 | 0.24505 | −0.29506 | 0.70885 | 0.844 | 28 | 0.406 |
| Pair 8 | F2–F1 | 1.58621 | 1.95516 | 0.36306 | 0.84250 | 2.32991 | 4.369 | 28 | 0.000 |
| Pair 9 | G2–G1 | 0.17241 | 1.51349 | 0.28105 | −0.40329 | 0.74811 | 0.613 | 28 | 0.545 |
| Pair 10 | H2–H1 | −0.86207 | 1.12517 | 0.20894 | −1.29006 | −0.43408 | −4.126 | 28 | 0.000 |
| Pair 11 | I2–I1 | −0.03448 | 1.40109 | 0.26018 | −0.56743 | 0.49846 | −0.133 | 28 | 0.896 |
| Pair 12 | J2–J1 | 0.65517 | 1.69613 | 0.31496 | 0.01000 | 1.30034 | 2.080 | 28 | 0.047 |
| Pair 13 | K2–K1 | 0.72414 | 1.19213 | 0.22137 | 0.27067 | 1.17760 | 3.271 | 28 | 0.003 |
| Pair 14 | L2–L1 | 0.24138 | 1.68301 | 0.31253 | −0.39880 | 0.88156 | 0.772 | 28 | 0.446 |
| Pair 15 | M2–M1 | 1.37931 | 1.44948 | 0.26916 | 0.82796 | 1.93066 | 5.124 | 28 | 0.000 |
| Pair 16 | N2–N1 | −0.89655 | 0.97632 | 0.18130 | −1.26792 | −0.52518 | −4.945 | 28 | 0.000 |
| Pair 17 | O2–O1 | 0.72414 | 0.99630 | 0.18501 | 0.34517 | 1.10311 | 3.914 | 28 | 0.001 |
| Pair 18 | P2–P1 | −0.31034 | 1.51430 | 0.28120 | −0.88635 | 0.26566 | −1.104 | 28 | 0.279 |

(continued)

**Table 4** (continued)

| | | Post-survey and pre-survey paired differences | | | | | t | df | Sig. (2-tailed) |
|---|---|---|---|---|---|---|---|---|---|
| | | Mean | Std. deviation | Std. error mean | 95% confidence interval of the difference | | | | |
| | | | | | Lower | Upper | | | |
| Pair 19 | Q2–Q1 | 1.34483 | 0.93640 | 0.17389 | 0.98864 | 1.70102 | 7.734 | 28 | 0.000 |
| Pair 20 | R2–R1 | 0.58621 | 1.78320 | 0.33113 | −0.09209 | 1.26450 | 1.770 | 28 | 0.088 |
| Pair 21 | S2–S1 | −0.27586 | 1.53289 | 0.28465 | −0.85894 | 0.30722 | −0.969 | 28 | 0.341 |
| Pair 22 | T2–T1 | −0.31034 | 0.84951 | 0.15775 | −0.63348 | 0.01279 | −1.967 | 28 | 0.059 |
| Pair 23 | U2–U1 | 0.44828 | 1.40372 | 0.26067 | −0.08567 | 0.98222 | 1.720 | 28 | 0.097 |
| Pair 24 | V2–V1 | 1.10345 | 2.16044 | 0.40118 | 0.28166 | 1.92523 | 2.750 | 28 | 0.010 |
| Pair 25 | W2–W1 | −0.10345 | 0.85960 | 0.15962 | −0.43042 | 0.22353 | −0.648 | 28 | 0.522 |
| Pair 26 | X2–X1 | 0.27586 | 1.03152 | 0.19155 | −0.11651 | 0.66823 | 1.440 | 28 | 0.161 |
| Pair 27 | Y1–Y2 | −0.55172 | 0.78314 | 0.14542 | −0.84961 | −0.25384 | −3.794 | 28 | 0.001 |

teachers have already established into classroom practice (Hubbard, 2008). In contrast, more recent efforts seem geared to helping teachers understand the educational potential of emerging AI technologies (Pokrivčáková, 2019). Pokrivčáková (2019) emphasizes that language teachers "do not need to be programmers or AI experts" to meaningfully incorporate chatbots into instruction (p. 146). Yet, the continuously developing educational technologies may place teachers into more sophisticated contexts that require advanced, rather than only basic, technological applications.

Confidence seems to be an essential aspect impacting how teachers are willing to use AI tools (Dai et al., 2020; Du et al., 2024; Ifenthaler et al., 2024; Tondeur et al., 2025). Some participants in the present study voiced concerns that developing chatbots would involve complex programming. However, many participants also said that learning a few basic concepts and creating a simple chatbot was much easier than they had anticipated. These findings, like those of Pokrivčáková (2019) and Liu (2025), seem to indicate that hands-on trainings aimed at boosting self-efficacy are likely to encourage AI technology integration in teaching. Individual confidence levels, however, are likely to affect the integration of technology. Some teachers may still require assistance in mastering the use of chatbots. It is likely that small incremental improvements in confidence will encourage the use of other educational technology.

## 4.2 How the Likelihood of Accepting the Chatbot Changed

Table 5 shows how different factors in the adapted theoretical framework differ before and after training. Generally, the training would make participants get more insights into the chatbot, feel easier to use, more socially influential, more enjoyable and better fitted to their usage, and more potential to apply into different teaching activities. Most factors show significant correlation with behavior intention ($p < 0.05$) and considerable changes, indicating that the training would significantly increase the likelihood of employing the chatbot in practice. After participants underwent training, the participants were also more aware of necessary support they would need in order to get deeper training and implementation in their teaching contexts.

**Perceiving risk** is the only factor observed as insignificantly correlated to behavior intention ($p > 0.05$) in both pre-survey and post-survey. It can also be inferred that the participants did not show much concern on the chatbot's reliability, accuracy, privacy and the risk of disruptive class engagement. The interview data also revealed more attention on usability, efficiency and convenience rather than any special recognized

**Table 5** Pearson *correlation analysis before and after training*

| | | Pre-survey | Post-survey |
|---|---|---|---|
| PE | Correlation | 0.596** | 0.793** |
| | Sig. (2-tailed) | 0.001 | 0.000 |
| EE | Correlation | 0.398* | 0.634** |
| | Sig. (2-tailed) | 0.040 | 0.000 |
| SI | Correlation | 0.485* | 0.769** |
| | Sig. (2-tailed) | 0.010 | 0.000 |
| FC | Correlation | 0.207 | 0.673** |
| | Sig. (2-tailed) | 0.299 | 0.000 |
| HM | Correlation | 0.640** | 0.836** |
| | Sig. (2-tailed) | 0.000 | 0.000 |
| HA | Correlation | 0.345 | 0.781** |
| | Sig. (2-tailed) | 0.078 | 0.000 |
| PR | Correlation | 0.108 | −0.248 |
| | Sig. (2-tailed) | 0.592 | 0.212 |
| SE | Correlation | 0.386* | 0.751** |
| | Sig. (2-tailed) | 0.047 | 0.000 |

**. Correlation is significant at the 0.01 level (2-tailed);
*. Correlation is significant at the 0.05 level (2-tailed)

risk or possible negative outcomes of the AI-developed chat. The most common worry among the participants was that they felt uncertain about "different ways of expression when someone asks and how to predict and prepare for unforeseen questions" (N. 26) and "a short period given to provide a truly interactive chatbot is impossible" (N. 7).

**PE**: Although some participants reported that the training was not as interesting as they expected, they also agreed with almost other participants that the training made them perceive chatbots more helpful for their teaching. Among the factors, PE has been identified as the most influential one to behavior intention (Morosan & DeFranco, 2016; Raman & Don, 2013; Venkatesh et al., 2012). Accordingly, correlations between the 2 factors were observed as strongly high and significant in both pre-survey ($r = 0.596$) and post-survey ($r = 0.793$). The qualitative data also shows PE made participants more likely to use the chatbot. "I'm quite satisfied as now I know how a chatbot operates, I have more ideas on how organize learning activities with chatbot for my students" (N. 21). Many other participants also highly appreciated the chatbot for its quick response, friendly user interface, personalized learning experience, flexibility with language choices and availability of media resources.

**EE**: Developing a chatbot did not require much effort and specialized background as the participants had supposed. That's why the correlation with BE increased from 0.398 to 0.634 in the post-survey. Positive correlation between EE and BI was noted by Raman and Don (2013), Zhou et al. (2010), and Venkatesh et al. (2012). The correlation significant of this study was also observed in both surveys. However, at the beginning there were some levels of uncertainty among the participants about skills and knowledge to be eligible for training participation. This possibility is also recommended by AlAwadhi and Morris (2008) who found that the EE can be moderated by prior technology experience and background. Some of the trainees did meet technical issues such as not being familiar with handling the functioning features, having problems with saving and editing content and connecting with the resources provided. Data loss and technical glitches were also sometimes reported. For example, as participant coded number 1 reported: "I completed the chatbot last Friday, saved and felt pleased with my product, but next morning when I started to submit, I realized that some parts of my work were lost without any reason". Nevertheless, most of the trainees learnt the chatbot very quickly after 3 h training and finally developed their own chatbot individually without much support within a week.

**SI**: After training, assumption on expectation of important surrounding social groups such as parents, friends, students, classmates, and/or colleagues played much stronger role (from 0.485 (pre-survey) to 0.769 (post-survey)) in predicting participants' intention to use the chatbot. In believing that people they valued preferred them to learn and apply chatbot in teaching, the participants are more inclined to employ the chatbot. This finding seems to contradict to a note of Phe et al. (2025) who also adopted UTAUT2 framework and found that Vietnamese learners of English did not show statistical significance between SI and BI. Phe et al. (2025), however, studied how undergraduate students evaluated a blended writing course. Participants in this study are teachers; thus, they are more experienced and may be more aware of attitudes of surrounding people. Such experience and awareness would make them believe more strongly in the effectiveness of using chatbot for teaching. As criticized by a participant having taught in a private school for 4 years, "finally, I support the idea of employing chatbot in suitable teaching activities. This chatbot will make the lessons more interesting and interactive to students. By employing this chatbot, a teacher is also employing a helpful assistant. The students can also have another 24 h partner to practice English with" (N. 8).

**FC**: In realizing that commanding a simple chatbot is not as complicated as having perceived, and that trainers can give in-time supports, there was a noticeable shift from non-significant FC-BI correlation of 0.207 (pre-survey) to a strong and significant correlation of 0.673 (post-survey). The participants come from different educational contexts with varied levels of available facilities for employing technologies to assist teaching, but in the post-survey they believed strongly in the possibility to organize such similar training and implement the chatbot in their teaching context.

**HM**: According to Alruwaili et al. (2025) and Collie and Martin (2024), teachers' motivation to incorporate AI tools is often shaped by the perceived potential for enhancing learning outcomes and increasing student engagement. The results showed that participants felt more motivated as they thought using the chatbot being trained for teaching is entertaining, enjoyable and convenient for both teacher and student. The correlation between HM with BI in this training experiment, which increased considerably from 0.640 (pre-survey) to 0.836 (post-survey), signifies that the training made the chatbot trainees more inclined to employ it.

**HA**: In the two surveys, the teachers were asked whether they felt like and were using Chatbot to assist their teaching. The correlation with BI moved from not significant ($p = 0.078$, $r = 0.345$) to highly significant ($p = 0.000$, $r = 0.781$) after the training experiment. Interestingly, as confirmed and explained by some participants, not all of the changes were directly from training, but from the increase in their confidence when they were asked. "I had no idea about kinds of chatbot I would be trained at the beginning, now I know that I have also used similar tools in my teaching, … especially when giving self-study tasks" (N. 23). Another participant noted that "using chatbots in teaching is quite familiar to me already, I usually allow students to use any chatbots they trust, from now I will also work myself to develop this chatbot, … I will try to develop some more chats with this given account" (N. 7).

**SE**: Inadequate technical training is a crucial factor to limit teachers' BI to adopt AI technology and increase the complexity of technology adoption (Alwaqdani, 2024; Ifenthaler et al., 2024; Pokrivčáková, 2019). Therefore, appropriate training will help to enhance self-efficacy and confidence of teachers in using chatbots. When self-efficacy changes, intention and actual integration of AI tools will also become more or less predictable (Li et al., 2022). In this study, SE's correlation with BI increased substantially from 0.386 (pre-survey) to 0.751 (post-survey). Accordingly, the training significantly boosted participants' perceived competence in commanding the trained chatbot, which in turn strongly increased their intention to integrate and sustain its use.

To examine more details into individual changes of acceptance before and after training, Cohen's *d* for effect size estimation was also calculated for the 27 participants. As shown in Table 6, large effects ($|d| \geq 0.5$) were observed in 12/27 participants, most of which are positive changes. They are teachers numbered as 4, 5, 6, 8, 10, 13, 15, 16, 17, 19, 21, and 27. Small to medium effects were observed with changes with teachers numbered as 2, 12, 20, 22, and 23. Meanwhile, Cohen's d shows small changes in the rest participants. The results demonstrate that the targeted training, measured by UTAUT2 framework (Venkatesh et al., 2012), had a meaningful impact on the participants' acceptance of the AI chatbot. Larger percentage of positive changes indicate that the training effectively enhanced students' confidence and acceptance level and also suggest that well-designed training interventions can significantly increase users' intention to adopt new technologies.

In conclusion, in studying the correlations between the constructs and analysing the qualitative feedback from the trainees, results indicate most of the constructs, except PR, serve as significant predictors of teachers' behavioral intention towards the chatbot. These findings are significant because they indicate not only the effectiveness of training, but how training shaped teachers' intention of AI adoption. The finding highly noted on the role of sufficient training on transforming teachers' beliefs and confidence in using new technologies that they may not have previous technical background or much experience with.

## 5 Conclusion and Implementation

This study has found some significant discoveries in answering its objectives and research questions. Most teachers demonstrate an increase in confidence and acceptance intention after the training session and coaching period. However, the mixed results also showed that some teachers exhibited self-confidence and acceptance levels declines. The declines are mainly due to technological intricacies, while the increases would come from the participants' gained skills and knowledge. The Pearson Correlation across different factors in the UTAUT2 framework demonstrated a behavioral shift among the participants. All factors show significant correlations between the pre-survey and post-survey except for PR factor, indicating that teachers have a more positive view on Chatbot after the training, seeing it as a practical and useful tool for their teaching. Most teachers found chatbot development less difficult than expected. They also demonstrated a shift from insignificant to significant correlation in the FC factor. These findings suggest that with appropriate and sufficient training and support, teachers have great potential to be fully capable of commanding such new technologies.

Findings from the study have led to some implications. It contributes theoretically by expanding UTAUT2 to examine EFL teachers' disengagement with AI chatbots. It has also shown that brief targeted training can significantly improve teachers' SE

**Table 6** Cohen's d calculation of effect size

| Participants | Comparison | t | Cohen's *d* | Interpretation |
|---|---|---|---|---|
| 1 | A2A–A1A | −0.297 | −0.06 | Small |
| 2 | B2A–B1A | 2.117 | 0.41 | Small–medium |
| 3 | A2–A1 | 0.361 | 0.07 | Small |
| 4 | B2–B1 | −6.282 | −1.21 | Large |
| 5 | C2–C1 | 4.683 | 0.90 | Large |
| 6 | D2–D1 | −4.700 | −0.90 | Large |
| 7 | E2–E1 | 0.844 | 0.16 | Small |
| 8 | F2–F1 | 4.369 | 0.84 | Large |
| 9 | G2–G1 | 0.613 | 0.12 | Small |
| 10 | H2–H1 | −4.126 | −0.79 | Large |
| 11 | I2–I1 | −0.133 | −0.03 | Small |
| 12 | J2–J1 | 2.080 | 0.40 | Small–medium |
| 13 | K2–K1 | 3.271 | 0.63 | Medium–large |
| 14 | L2–L1 | 0.772 | 0.15 | Small |
| 15 | M2–M1 | 5.124 | 0.99 | Large |
| 16 | N2–N1 | −4.945 | −0.95 | Large |
| 17 | 02–01 | 3.914 | 0.75 | Large |
| 18 | P2–P1 | −1.104 | −0.21 | Small |
| 19 | Q2–Q1 | 7.734 | 1.49 | Very large |
| 20 | R2–R1 | 1.770 | 0.34 | Small–medium |
| 21 | S2–S1 | −0.969 | −0.19 | Small |
| 22 | T2–T1 | −1.967 | −0.38 | Small–medium |
| 23 | U2–U1 | 1.720 | 0.33 | Small–medium |
| 24 | V2–V1 | 2.750 | 0.53 | Medium |
| 25 | W2–W1 | −0.648 | −0.12 | Small |
| 26 | X2–X1 | 1.440 | 0.28 | Small |
| 27 | Y2–Y1 | 3.794 | 0.73 | Large |

and confidence. On a practical level, the study illustrates how even novice practitioners, with the right scaffolding and a motivating educational environment, can gain considerably greater technological proficiency. It is suggested that courses and training about commanding new technologies should be made available and official for educators. Further guidance and up-to-date facilities should also be provided by institutions for long-term effectiveness. Moreover, EdTech companies may refer to the insights of this study to develop products which empower teachers to have a more active role in the process of creating and using new educational technologies.

However, the study has a number of weaknesses. The three-week duration of the study, the reliance on self-reported confidence rather than assessing performance directly, and the small sample of 27 teachers from a single school all limit the study's potential for broader dissemination. These factors constrain the extent to which conclusions can be extended to broader teacher populations.

Following this study, longitudinal research should establish whether increases in confidence can be maintained across long intervals and whether this confidence is actually reflected in the use of chatbots during teaching practices. It would also be helpful to conduct studies across a greater number of institutions and with larger samples. This extension may help disentangle the contextual variables of confidence that include administrative support, digital literacy, and subsequent training.

**Acknowledgement of AI Assistance** ChatGPT was employed to assist in searching for relevant and downloadable literature. The similarity and AI match rates were checked by Turnitin.

**Conflict of Interest** There are no conflicts of interest related to this work among the authors.
**Author Verification** The study is original and has not been published elsewhere or is under consideration for publication. All participants consented.

## Appendix

The adapted UTAUT2 (Venkatesh et al., 2012) for chatbot acceptance survey.

| | |
|---|---|
| Performance expectancy | • I believe that Chatbot is helpful for my teaching<br>• The use of a Chatbot would increase my chances of simulating classroom teaching<br>• Using a Chatbot can improve my teaching practice ability<br>• Using a Chatbot can make me teach more efficiently |
| Effort expectancy | • It is easy for me to learn how use a Chatbot<br>• I can easily plan teaching activities using a Chatbot<br>• I can quickly learn to use a Chatbot for teaching |
| Social influence | • People who are important to me (such as parents, friends, students, classmates, AND/OR colleagues, etc.) believe I should learn and use Chatbot for teaching<br>• People who affect my behavior (such as parents, friends, students, classmates, AND/OR colleagues, etc.) believe I should learn and use Chatbot for teaching<br>• People I value prefer me to learn and use Chatbot for teaching |
| Facilitating conditions | • My workplace can provide Chatbot training and teaching equipment<br>• My workplace will provide training in using the Chatbot for us<br>• I can get help from others when I have difficulties using a Chatbot for teaching |
| Hedonic motivation | • Using Chatbot for teaching is fun/enjoyable and entertaining<br>• Using Chatbot for teaching will make me happy with my lessons |

| | |
|---|---|
| | • Using Chatbot for teaching will make my students pleased |
| Habit | • Currently, using of Chatbot for teaching is my habit<br>• Currently, using Chatbot for teaching is natural to me<br>• After using chatbot for teaching, I will like using the chatbot |
| Perceived risks | • The (coming) training of Chatbot will not as good as we expected<br>• I feel nervous with the Chatbot training<br>• I can't guarantee my privacy security when learning and using the Chatbot |
| Self-efficacy | • With a Chatbot, I have the confidence to complete my teaching tasks<br>• I can solve the difficulties encountered in the Chatbot training<br>• I believe that with determination, I can successfully use a Chatbot for teaching |
| Behavior intention | • I want to use Chatbot to practice my teachers' professional skills<br>• If the effect is good, I am willing to recommend the Chatbot training to others<br>• In the future, I plan to study higher courses offered about Chatbot<br>• I will develop a Chatbot and use as a part of my students' homework |

## References


AlAwadhi, S., & Morris, A. (2008). The use of the UTAUT model in the adoption of e-government services in Kuwait. In *Proceedings of the 41st annual Hawaii international conference on system sciences (HICSS 2008)* (pp. 219–219). IEEE. https://doi.org/10.1109/hicss.2008.452

Alruwaili, A. R., Kianfar, Z., & Noughabi, M. A. (2025). Exploring EFL teachers' motivation and influential factors in adopting AI for language teaching. *Language Related Research, 16*(5), 93–124. https://doi.org/10.48311/LRR.16.5.4

Alwaqdani, M. (2024). Investigating teachers' perceptions of artificial intelligence tools in education: Potential and difficulties. *Education and Information Technologies*. https://doi.org/10.1007/s10639-024-12903-9

Amemasor, S. K., Oppong, S. O., Ghansah, B., Benuwa, B. B., & Essel, D. D. (2025). A systematic review on the impact of teacher professional development on digital instructional integration and teaching practices. *Frontiers in Education, 10*, 541031. https://doi.org/10.3389/feduc.2025.1541031

Amer jid Almahri, F. A., Bell, D., & Gulzar, Z. (2024). Chatbot technology use and acceptance using educational personas. *Informatics, 11*(2), 38. https://doi.org/10.3390/informatics11020038

Bandura, A. (1994). Self-efficacy. In V. S. Ramachaudran (Ed.), *Encyclopedia of human behavior* (Vol. 4, pp. 71–81). http://www.des.emory.edu/mfp/BanEncy.html

Bandura, A. (1997). *Social learning theory*. General Learning Press.

Buchner, J., & Hofmann, M. (2022). The more the better? Comparing two SQD-based learning designs in a teacher training on augmented and virtual reality. *International Journal of Educational Technology in Higher Education, 19*(1), 24. https://doi.org/10.1186/s41239-022-00300-1

Chau, H. T. H., Bui, H. P., & Dinh, Q. T. H. (2024). Impacts of online collaborative learning on students' intercultural communication apprehension and intercultural communicative competence. *Education and Information Technologies, 29*, 7447–7464. https://doi.org/10.1007/s10639-023-12100-0

Chen, L., Chen, P., & Lin, Z. (2020). Artificial intelligence in education: A review. *IEEE Access, 8*, 75264–75278. https://doi.org/10.1109/access.2020.2988510

Collie, R. J., & Martin, A. J. (2024). Teachers' motivation and engagement to harness generative AI for teaching and learning: The role of contextual, occupational, and background factors. *Computers and Education: Artificial Intelligence, 6*, 100224. https://doi.org/10.1016/j.caeai.2024.100224

Dai, Y., Chai, C. S., Lin, P. Y., Jong, M. S. Y., Guo, Y., & Qin, J. (2020). Promoting students' well-being by developing their readiness for the artificial intelligence age. *Sustainability, 12*(16), 6597. https://doi.org/10.3390/su12166597

Ding, D., & Yusof, M. B. (2025). Investigating the role of AI-powered conversation bots in enhancing L2 speaking skills and reducing speaking anxiety: A mixed methods study. *Humanities and Social Sciences Communications, 12*(1), 1–16. https://doi.org/10.1057/s41599-025-05550-z

Dörnyei, Z., & Taguchi, T. (2009). *Questionnaires in second language research: Construction, administration, and processing*. Routledge. https://doi.org/10.4324/9781003331926

Du, H., Sun, Y., Jiang, H., Islam, A. Y., & Gu, X. (2024). Exploring the effects of AI literacy in teacher learning: An empirical study. *Humanities and Social Sciences Communications, 11*(1). https://doi.org/10.1057/s41599-024-03101-6

Du, J., & Daniel, B. K. (2024). Transforming language education: A systematic review of AI-powered chatbots for English as a foreign language speaking practice. *Computers & Education: Artificial Intelligence, 6*, 100230. https://doi.org/10.1016/j.caeai.2024.100230

Hirsch, S. E., Ely, E., Lloyd, J. W., & Isley, D. (2018). Targeted professional development: A data-driven approach to identifying educators' needs. *School-University Partnerships, 11*(2), 84–91. https://files.eric.ed.gov/fulltext/EJ1199818.pdf

Hsu, Y. Y., & Lin, C. H. (2020). Evaluating the effectiveness of a preservice teacher technology training module incorporating SQD strategies. *International Journal of Educational Technology in Higher Education, 17*(1), 31. https://doi.org/10.1186/s41239-020-00205-2

Huang, J., Saleh, S., & Liu, Y. (2021). A review on artificial intelligence in education. *Academic Journal of Interdisciplinary Studies, 10*(3), 206. https://doi.org/10.36941/ajis-2021-0077

Hubbard, P. (2008). CALL and the future of language teacher education. *CALICO Journal, 25*(2), 175–188. https://doi.org/10.1558/cj.v25i2.175-188

Ifenthaler, D., Majumdar, R., Gorissen, P., Judge, M., Mishra, S., Raffaghelli, J., & Shimada, A. (2024). Artificial intelligence in education: Implications for policymakers, researchers, and practitioners. *Technology Knowledge and Learning*. https://doi.org/10.1007/s10758-024-09747-0

Kent, A. M., & Giles, R. M. (2017). Preservice teachers' technology self-efficacy. *SRATE Journal, 26*(1), 9–20. https://files.eric.ed.gov/fulltext/EJ1134392.pdf

Kim, A., & Su, Y. (2024). How implementing an AI chatbot impacts Korean as a foreign language learners' willingness to communicate in Korean. *System, 122*, 103256. https://doi.org/10.1016/j.system.2024.103256

Koç, F. Ş., & Savaş, P. (2025). The use of artificially intelligent chatbots in English language learning: A systematic meta-synthesis study of articles published between 2010 and 2024. *ReCALL, 37*(1), 4–21. https://doi.org/10.1017/S0958344024000168

Labadze, L., Grigolia, M., & Machaidze, L. (2023). Role of AI chatbots in education: Systematic literature review. *International Journal of Educational Technology in Higher Education, 20*(1), 56. https://doi.org/10.1186/s41239-023-00426-1

Li, B., Sun, J., & Oubibi, M. (2022). The acceptance behavior of blended learning in secondary vocational school students: Based on the modified UTAUT model. *Sustainability, 14*, 1–16. https://doi.org/10.3390/su142315897

Limayem, M., Hirt, S. G., & Cheung, C. M. (2007). How habit limits the predictive power of intention: The case of information systems continuance. *MIS Quarterly*, 705–737. https://doi.org/10.2307/25148817

Liu, N. (2025). Exploring the factors influencing the adoption of artificial intelligence technology by university teachers: The mediating role of confidence and AI readiness. *BMC Psychology, 13*(1), 311. https://doi.org/10.1186/s40359-025-02620-4

Moghavvemi, S. (2015). Impact of perceived self-efficacy and capability to use IT innovation on individual use behaviour. SSRN 2561739. https://doi.org/10.2139/ssrn.2561739

Morosan, C., & DeFranco, A. (2016). It's about time: Revisiting UTAUT2 to examine consumers' intentions to use NFC mobile payments in hotels. *International Journal of Hospitality Management, 53*, 17–29. https://doi.org/10.1016/j.ijhm.2015.11.003

Nguyen, L. T., Bui, H. P., & Ha, X. V. (2024). Scaffolding in genre-based L2 writing classes: Vietnamese EFL teachers' beliefs and practices. *International Journal of Applied Lingusitics in Language Teaching, 63*(2), 1275–1293. https://doi.org/10.1515/iral-2023-0125

Panagiotidis, P. (2024). LLM-based chatbots in language learning. *European Journal of Education (EJED), 7*(1), 102–123. https://files.eric.ed.gov/fulltext/EJ1455392.pdf

Phe, C. Q., Chowdhury, R., & Thu, L. T. T. (2025). Using the UTAUT2 model to determine the factors affecting students' acceptance of blended learning for English writing. *CALL-EJ, 26*(3), 24–42. https://doi.org/10.54855/callej.252632

Pokrivčáková, S. (2019). Preparing teachers for the application of AI-powered technologies in foreign language education. *Journal of Language and Cultural Education, 7*(3), 135–153. https://doi.org/10.2478/jolace-2019-0025

Raman, A., & Don, Y. (2013). Preservice teachers' acceptance of learning management software: An application of the UTAUT2 model. *International Education Studies, 6*(7), 157–164. https://doi.org/10.5539/ies.v6n7p157

Riazi, A. M. (2017). *Mixed methods research in language teaching and learning*. Equinox Publication.

Tang, W., Chaiyasoonthorn, W., & Chaveesuk, S. (2024). Understanding the impact of self-efficacy, personal innovativeness and content quality on mobile learning usage in higher vocational colleges. *Espacio, Tiempo y Educación, 11*(1), 146–164. https://doi.org/10.14516/ete.11110

Tin, T. N., & Bui, H. P. (2024). Sampling, generalizability, and reliability. In H. P. Bui (Ed.), *Applied linguistics and language education research methods: Fundamentals and innovations* (pp. 103–121). IGI Global.

Tondeur, J., Trevisan, O., Howard, S. K., & Van Braak, J. (2025). Preparing preservice teachers to teach with digital technologies: An update of effective SQD-strategies. *Computers & Education, 232*, 105262. https://doi.org/10.1016/j.compedu.2025.105262

UNESCO. (2023). Technology in education: A tool on whose terms? In *Global education monitoring report 2023, summary*. UNESCO, Paris, and SEAMEO, Bangkok. https://unesdoc.unesco.org/

Venkatesh, V., Thong, J. Y. L., & Xu, X. (2012). Consumer acceptance and use of information technology: Extending the unified theory of acceptance and use of technology (UTAUT2). *MIS Quarterly, 36*(1), 157–178. https://doi.org/10.2307/41410412

Wang, C., Zou, B., Du, Y., & Wang, Z. (2024). The impact of different conversational generative AI chatbots on EFL learners: An analysis of willingness to communicate, foreign language speaking anxiety, and self-perceived communicative competence. *System, 127*, 103533. https://doi.org/10.1016/j.system.2024.103533

Wollny, S., Schneider, J., Di Mitri, D., Weidlich, J., Rittberger, M., & Drachsler, H. (2021). Are we there yet?—A systematic literature review on chatbots in education. *Frontiers in Artificial Intelligence, 4*, 654924. https://doi.org/10.3389/frai.2021.654924

Yan, M., Castro, P., Cheng, P., & Ishakian, V. (2016). Building a chatbot with serverless computing. In *Proceedings of the 1st international workshop on mashups of things and APIs* (pp. 1–4). https://doi.org/10.1145/3007203.3007217

Zhou, T., Lu, Y., & Wang, B. (2010). Integrating TTF and UTAUT to explain mobile banking user adoption. *Computers in Human Behavior, 26*(4), 760–767. https://doi.org/10.1016/j.chb.2010.01.013

**Dr. Nguyen Truong Sa** obtained his Doctor of Education Degree specifying in English language teaching from Leicester University. He is now the Dean of Faculty of Foreign Languages, Industrial University of Ho Chi Minh City, Vietnam. His publication are related to applied technology in SLA and language teaching.

**Nguyen Thi Diem Thi** is a lecturer at Faculty of Foreign Languages, Industrial University of Ho Chi Minh City (IUH). She is currently pursuing her Ph.D. at HUFLIS, Hue University. Her research interests include sociocultural theory, activity theory, mobile/technology-assisted and autonomous/self-directed/self-regulated language learning, and the use of artificial intelligence in language education.

**Ho Thi Hien** is a lecturer of English at the Industrial University of Ho Chi Minh city, Vietnam. She is a Ph.D. student in TESOL at HUFLIS, Hue University. Her research interests include Information and Communication Technology (ICT) in English language teaching, chatbot-human interaction, learners' satisfaction, and Intercultural Communicative Competence (ICC).

**Hoang Thanh Dung** is a teaching assistant at the Faculty of Foreign Languages, Industrial University of Ho Chi Minh City (IUH). She got her Master Degree in TESOL at University of Social Sciences and Humanity, Ho Chi Minh City.

# Building Audiobooks with AI Support as Digital Resources for ESL Learning

**Phan Thi Thanh Thao**

**Abstract** In today's technological era, second language learning has been enriched by a wide range of digital resources, many of which are strongly influenced by Artificial Intelligence (AI). Using AI tools to build audiobooks for learning English as a Second Language (ESL) shows an innovative educational strategy that combines technological advancement with language learning. This book chapter reveals some benefits of using audiobooks in learning English language skills, especially listening and speaking skills, and practicing pronunciation and developing vocabulary. In addition, the use of audiobooks provides more chances for ESL learners to access authentic materials in real communication. However, many English audiobooks are not available, costly and unsuitable for learners' English competence levels. Hence, it is necessary to design audiobooks that can help ESL learners engage with authentic resources. Based on an extensive literature review on AI's role in creating digital resources for ESL learning, and audiobooks' issues in ESL learning, this study focuses on building audiobooks with AI tools. An experiment was conducted at a university in Vietnam with and without using audio texts generated by AI in ESL classes of listening skill practice. The experimental results shown by students' scores of English listening tests and their feedback indicate the necessity of providing some practical tasks of creating audiobooks for the development of authentic resources and technology application into ESL learning. Finally, some pedagogical implications on ESL learner and teacher-training in the present digital era have been discussed in this study.

**Keywords** Digital resources · AI tool · Audiobooks · ESL learners · Listening skills

P. T. T. Thao (✉)
University of Foreign Languages and International Studies, Hue University, Hue, Vietnam
e-mail: pttthao@huflis.edu.vn

 
H. P. Bui and L. P. H. Huong (eds.), *Artificial Intelligence in Education and Academic Research*, Studies in Computational Intelligence 1269,
https://doi.org/10.1007/978-3-032-18392-7_5

## 1 Introduction

In the world of rapidly developing advanced technology, the rise of Artificial Intelligence (AI) has brought significant changes and increased efficiency and accuracy in various aspects of life including healthcare, transportation, finance, and education (Chen, 2024; Geraldin et al., 2022; Khalilova et al., 2025; Kothai, 2024). In the education aspect, AI can enhance education by bringing knowledge to more people, inspiring creativity, summarizing existing materials, creating more authentic materials for language learners and teachers (Handayan et al., 2025). For instance, AI boosts practical abilities of learning language skills like writing and offers a trustworthy simulation dialogue platform for spoken English, which makes learning English simpler and more interesting (Zanyar et al., 2023). In addition, the integration of artificial intelligence into language learning holds significant potential to transform new pedagogical approaches in the use of digital resources. In fact, the use of AI in language learning has the potential to revolutionize the way people learn languages, which has initiated a change in thinking, promising more tailored, efficient, and engaging learning experiences (Chen, 2024; Krylov, 2023), for instance, AI tools can be applied into developing digital resources for learning English as a second language such as e-books, e-lectures, audiobooks, etc.

It is apparent that digital books give valuable benefits to ESL learners such as authentic language resources to improve their listening skills, reading comprehension and speaking fluency (Serafini, 2004; Tran et al., 2021). However, many people are still coping with challenges in accessing suitable digital resources because many existing materials are often costly, inappropriate for different levels of language competence. These barriers can limit students' ability to fully engage with authentic English resources and hinder their overall language development (Javad & Leila, 2015). Therefore, this study introduces the advantages of building audiobooks for ESL learners with AI tools' support. It also demonstrates how AI can facilitate the creation of audiobooks that meet ESL learners' needs of improving language and technology knowledge.

## 2 The Role of AI in Creating Digital Resources

The term *Artificial Intelligence* (AI) has been defined as the science that uses machines to do things which require human intelligence (Fjelland, 2020). This definition can be understood in various fields that aim to imitate human behavior via technology or methods in our real world (Batista et al., 2024; Farrelly & Baker, 2023). In fact, AI has been popularly applied in designing, building, and providing ESL learners with digital resources that contain adaptive, interactive, and authentic content with remarkable efficiency (Ghafar et al., 2023).

In fact, one major area where AI has had a profound impact is content creation. Thanks to advancements and breakthroughs in generative language models, AI generated content adding to some conventional content creation methods such as user-generated content and professional-generated content, which has become a key focus in the world (Qian et al., 2025). AI systems can offer tailored digital resources, including audiobooks, podcasts, or reading passages, which might be used in a flexible way that meets individual needs, especially in learning English as a second language in a non-native English-speaking environment. It maximizes the teaching impact of ESL while increasing students' practice ability. Students may communicate in English more quickly since those digital resources are available and easy to use (Zanyar et al., 2023). Similarly, it is highlighted that the importance of using AI applications for developing students' English skills, particularly, in practicing listening skills. Employing AI applications helps develop students' English language skills and language communication skills through intelligent dialog boxes and other options depending on the users' purposes, for example, recording their voice or creating audiobooks (Mukhallafi, 2020).

For creating digital resources like audiobooks, AI known as an effective tool in the natural language processing domain (NLP), has a special function, namely text-to-speech (TTS), which enables learners to rapidly convert any written texts into spoken audio, making it possible to create audiobooks with realistic pronunciation, intonation, and pacing. This magnificent application of AI has offered the ESL learners diverse and valuable open resources, for example, audiobooks are read with multiple accents and speaking styles, which used to be hard to achieve with human-recorded materials alone in the past.

Since these tools have a high ability to simulate diverse accents of people at different ages, genders from various regions in the world to provide many kinds of speeches in communicative contexts, the digital resources like audiobooks are of great authenticity that helps learners approach the complex language use in the real world. ESL learners feel familiar and close to the spoken voice from audiobooks as if they were communicating with real people.

In summary, AI tools are believed to be effective in building digital resources for learning ESL. AI has brought ESL learners many opportunities for practicing language skills inside and outside classrooms with only an electronic device such as a smartphone, a tablet, or a computer. In addition, AI gives them a huge number of resources with the novelty in language education in the digital technology era.

## 3 Audiobooks in Learning English as a Second Language

### *3.1 What is an Audiobook?*

An audiobook is generally explained as an audio recording of a written document which is read or narrated. Audiobooks are distributed in digital formats, which can be used with a variety of technological devices such as smartphones, tablets, and computers (Sharma, 2025).

Audiobooks were first introduced in 1931 as part of the *Books for the Adult Blind Project*. Since then, their application has expanded considerably beyond the initial purpose. Early uses of audiobooks primarily focused on supporting children's reading comprehension (Koskinen et al., 2000) and assisting struggling readers (Harris, 2001; Stone-Harris, 2008). With the rapid advancements of the digital era, their development and accessibility have grown remarkably. Despite this growth, limited research has specifically examined the impact of audiobooks on the listening skills of ESL learners remains limited (Kartal & Simek, 2017). Addressing this gap, this study seeks to investigate the features and pedagogical benefits of audiobooks in the context of ESL learning.

## 3.2 Features of Audiobooks

Among the various features of audiobooks identified by linguists, educators, and practitioners, it is noticeable three main characteristics including popularity, authenticity, and flexibility/accessibility.

### 3.2.1 Popularity

Audiobooks have become a popular medium among ESL learners who want to improve their skills by listening to audio texts. In the United States, about 67 million American people listen to audiobooks each year, especially, a high rate of the younger tend to do that. It is reported that nearly 48% of audiobook listeners are under the age of 35. To meet this requirement, the audiobook industry was valued at approximately $900 million in 2009 and increased to $2.1 billion in 2016 (Rizal et al., 2022). Not only in English-speaking countries, the popularity of audiobooks is also widely increased in many nations where English is learnt as a second or foreign language. The expansion of this medium has been further facilitated by the rapid advances of the Fourth Industrial Revolution, during which learners can easily access digital platforms and online resources for both educational and professional purposes.

### 3.2.2 Authenticity

Audiobooks frequently provide authentic materials originally recorded for a general audience rather than specifically designed for language learners. In fact, authenticity is considered a foundation of the listening competence achievement. Since authentic material should be one of the main inputs for the learner, Gilakjani and Sabouri (2016) offered some useful suggestions for ESL student to overcome their difficulties in English listening comprehension by providing authentic listening materials helping them understand better the native speakers' speech. Similarly, to improve the ESL learners' listening skills, it is crucial to adapt and develop the listening materials that match the learners' interests and background and linguistic knowledge, it means that these authentic materials might offer the ESL learners better listening strategies as their learning inspiration and motivation has been increased in the classroom (Saraswaty, 2018).

In the adapting and improving listening materials, it is obvious that students differ in their learning styles and ability, therefore, teachers should adopt and adapt listening materials that match their students' interest and background since the listening materials only become Saraswaty (2018). Meanwhile, Ferris (1996) emphasizes that teachers often try to incorporate authentic input into their teaching practices, particularly by offering learners opportunities to engage in real-life lectures. Hamouda (2013) suggests one of the useful ways to enhance the ESL learners' listening competence. It means that students can use the topical and linguistic knowledge in authentic texts providing them with a better chance to understand the content they are going to listen to. In short, these issues highlight the authenticity of audiobooks as reliable resources for ESL learners when having a chance to access real speakers' voices in genuine communication contexts.

### 3.2.3 Flexibility and Accessibility

In addition to popularity and authenticity, audiobooks are well-known for their flexibility and accessibility, which makes them more proper for ESL learning. It is apparent that audiobooks can be accessed with various kinds of digital devices like smartphones, tablets, computers, and portable music players. This enables ESL learners to engage with language input in diverse contexts anywhere and anytime (Renandya & Farrell, 2012).

Besides, using audiobooks allows ESL learners to study in a flexible way inside or outside class, and improve remarkably their self-study. Especially, in listening skill practice, learners discover many functions of audiobook AI-generated tools such as pausing, replaying, or adjusting the playback speed based on their language competence levels. Such functions are especially valuable for those who are struggling with unfamiliar vocabulary, idiomatic expressions, or terminology, as they can listen to audiobooks many times to understand them (Chang & Millett, 2015).

In addition, it is easier to access audio books at digital libraries. In some cases, many audiobooks are accompanied by electronic or printed texts, which supply the learners with simultaneous listening and reading tasks. This enhances both comprehension and vocabulary retention (Woodall, 2010).

In brief, the key qualities mentioned above in audiobooks make them particularly useful in learning ESL. Due to the fact that most audiobooks are designed for common audience, they might give ESL learners conversations/dialogues, or stories spoken or read by real people with natural voice, pronunciation, rhythm, and intonation, thereby they might enhance their listening competence. As a result, audiobooks have become an effective resource with numerous benefits for ESL learners.

## 3.3 *Benefits of Using Audiobooks in ESL Learning*

The use of audiobooks in ESL learning offers a wide range of pedagogical benefits, including the development of listening comprehension skills and vocabulary, the enhancement of fluency in communication, increase in learning autonomy and motivation. They are the focus strength of language achievements that are extremely necessary for ESL learners.

### 3.3.1 Listening Comprehension Development

Some experimental studies found that listening to audiobooks improves both reading comprehension and listening skills (Jacobs, 2006; Kartal & Simek, 2017), but listening skills can be improved more greatly. In fact, audiobooks expose learners to authentic speech patterns, helping them understand natural spoken English with accurate pronunciation. Since pronunciation is extremely important in practice listening and speaking, most learners are likely to imitate the native speakers' accents and pronunciation who are often chosen as narrators of audiobooks. Thus, language learners do not feel timid about their pronunciation, and they can correct their pronunciation mistakes by listening and repeating as much as possible. Briefly, audio books have many positive effects on the pronunciation of EFL learners (Saka, 2015; Tagninezhad et al., 2015).

### 3.3.2 Increasing Vocabulary Retention

Audiobooks also play an important role in increasing the ESL learners' vocabulary retention. When learners listen to audiobooks in combination with printed or electronic texts, they can connect spoken and written forms of words, which enhances word recognition and promotes long-term retention (Woodall, 2010). It is pointed out three major differences between audiobooks and traditional reading aloud that make the ESL learners improve their vocabulary retention. First, whereas reading aloud is restricted in space and time, listening to audiobooks does not have such constraints (Kartal & Simek, 2017).

Therefore, ESL learners are able to memorize more vocabulary in a comfortable learning environment without time constraints and space. Another difference is related to voice. While reading aloud necessitates a face-to-face encounter, audiobooks can be listened to via computers, MP3 players, smartphones, CDs, cassettes, Walkman, and so on. Audiobooks releasing stories with an appealing voice make ESL learners excited and involved in the content full of vocabulary; as a result, their vocabulary knowledge or retention will be increased.

#### 3.3.3 Rising Fluence in Communication

In addition to listening and reading skills, which have improved for ESL learners when using audiobooks, their speaking skills also have opportunities to be enhanced. Since their repeated listening to audiobooks enables them to follow rhythm, intonation, and phrasing, they can figure out with ease how to improve speaking fluency. What is more, one of the advantages audiobooks is that stories or documents are often read aloud by professional narrators, actors or actresses, and even the authors (Wolfson, 2008). This motivates ESL learners to imitate and rehearse stories many times, which helps them in their significant speaking fluency.

#### 3.3.4 Learning Autonomy and Motivation

Another benefit lies in the ability to enhance the ESL learners' motivation and autonomy. The use of audiobooks in a flexible way allows them to self-study in various contexts, focusing on independent language learning. Moreover, the popularity and accessibility of audiobooks across digital platforms allows learners to select materials suitable for their interests and language targets both inside and outside the classroom (Renandya & Farrell, 2012). This can make a great contribution to ESL learners' autonomy and motivation. Listening to stories read by real famous people with soft sounds apparently encourages learners to focus on their language knowledge, which will be developed in the right way. Learners are also highly motivated to interact with other celebrities or idols whom they admire and respect.

In summary, with the benefits listed above, it is evident that using audiobooks can help increase learning outcomes and efficiency of technological application into ESL learning process.

### *3.4 Challenges of Using Audiobooks*

Everything has both positive and negative sides, and using audiobooks is not an exception. Despite many benefits, audiobooks also present several limitations that may reduce their effectiveness in ESL learning.

#### 3.4.1 Difficulties in Reading and Listening Comprehension Skills

It is acknowledged that audiobooks can develop both listening and reading skills in an advanced technological learning environment when they allow ESL learners to repeat documents until they understand it. However, some authentic audiobooks often use advanced vocabulary, a wide range of idiomatic expressions, and collocations, which are quite hard for beginners or those having a pre-intermediate level of language competence to understand (Chang, 2011). Those difficulties in the whole material comprehension due to some specialized terms or vocabulary, idioms, and expressions may reduce the ESL learners' focus on reading and listening skills. Furthermore, some learners relying solely on audiobooks may neglect visual reading practice, which leads to their lack of concentration on vocabulary and grammar knowledge (Wolfson, 2008) and may lower their motivation for ESL learning (Chang, 2011).

#### 3.4.2 Quality and Costly Audiobooks

Within the context of increasingly commercialized education in many countries around the world, audiobooks are often regarded as one of the marketable products that may not be freely available for people to access. This trend has significant implications for the growing population of ESL learners, many of whom may face financial barriers to obtaining high-quality and pedagogically valuable audiobooks. Some high-quality and valuable audiobooks are costly and require payment for users. Besides, not all ESL learners may have easy access to the available resources of audiobooks to improve their language learning.

#### 3.4.3 Limited Interaction

Most audiobooks are considered linear or non-interactive digital tools, which provide ESL learners with a lack of opportunities for interacting with their peers, instructors, or teachers. This limits the way of giving immediate feedback, questions for a deeper understanding of the material content (Rizal et al., 2022). Similarly, it is argued that audiobooks are inactive in comparison to print books since readers of audiobooks may obtain more passive experience, as they are restricted to simply listening to the story (Rizal et al., 2022).

In short, the use of audiobooks is facing some obstacles mentioned above even though the number of audiobooks and their users is steadily increasing. However, integrating audiobooks into teaching and the learning environment, especially into the ESL setting, is very important (Kartal & Simek, 2017). Hence, it is essential to build audiobooks with AI tools to enrich the authentic digital resources for both ESL learners and teachers.

## 4 Building Audiobooks with AI Tools

### *4.1 AI Tools Used to Create Audiobooks*

Technological development in recent years has arisen the appearance of numerous AI tools which can be used for audiobook production. In fact, a wide range of AI-powered text-to-speech (TTS) tools such as ElevenLabs, Speechify, Play HT,

WellSaid Labs, Murf AI, Dupdub AI, etc. are now widely applied to generate and convert written documents into audiobooks with natural sounding voices. In this study, some popular AI tools are introduced so that ESL teachers and learners could select the proper ones for their building and using of English audiobooks.

### 4.1.1 ElevenLabs

Among the current AI-based audiobook generators, ElevenLabs has been the top choice since it can offer a large number of realistic voices for various applications such as dubbing in 50 languages and voice cloning (with more than 1,200 AI voices). In addition, ElevenLabs platform offers a user-friendly interface which facilitates efficient voice generation while enabling users to conveniently access and utilize its advanced features.

### 4.1.2 Speechify

Speechify is another choice for beginners using text-to-speech technology or converting various documents into audio natural-sounding audios, because of its user-friendly interface. The designer of this AI tool, Cliff Weitzman, was diagnosed with dyslexia. To help other people having the same problem of accurate pronunciation, he decided to create this tool. Its main features are shown with a library of more than 200 AI voices in 50+ languages.

### 4.1.3 PlayHT

Being a top-rated AI audio generator, Play HT has more than 800 AI voice options in 142 languages. It is rather simple and easy to operate the PlayHT platform when the users just type or import their text, select a voice and generate an audio. Furthermore, they can adjust the appropriate speech, emotion, intonation, intensity, stability settings to create their speech.

### 4.1.4 WellSaid Labs

As a cutting-edge AI tool, WellSaid Labs can be simply used to build AI-generated audiobooks with more than 80 AI voices only in English with a wide range of accents from English native speaking countries such as Australia, the UK, Canada, Ireland, New Zealand, Scotland, South Africa, and the United States.

#### 4.1.5 Murf AI

In comparison with ElevenLabs and PlayHT, Murf AI is a popular text-to-speech AI voice generator with fewer voice options (more than 150 AI voices) and fewer languages (just 20 languages). The way of operating this AI tool is also simple by asking users to input their text, select a voice and generate an audio product. Interestingly, Murf AI can interpret punctuation and emotional cues with natural-sounding voiceovers.

#### 4.1.6 DupDub AI

DupDub is an AI-generated tool that offers distinguishing characteristics suited to the creation of audiobooks. This tool provides users with over 700 voiceovers in more than 90 languages. For learners and educators in second language settings, DupDub AI tool can significantly decline the time and cost of producing English audiobooks for ESL teaching and learning.

### *4.2 How to Build Audiobooks with Dupdub*

In this book chapter, DupDub AI tool (www.dupdub.com) has been chosen to generate audiobooks due to its advantages like being easy-to-use, multifunctional, free of charge and suitable for educational environment. Since getting the permission of Dupdub Support via email informing that researchers are allowed to use all the outputs created with Dupdub during 3 day free trial for academic and research purposes, we could use and share some images and audio texts generated by DupDub. This section presents main steps of creating English audiobooks which can be used as authentic materials in English listening skill practice at a university in Vietnam. Here are five main steps to create an English audiobook with DupDub AI tool.

**Step 1: Preparation for creating an audiobook**

To begin, the user should create an account with his/her email to sign in DupDub AI and complete his/her description. Figure 1 explores the user-friendly interface of this AI tool.

This tool provides a range of products including video, audio, text and others. In each product, for example, video, the user can select different functions like AI Avatar, video translation, video editing, video downloader and video recording which depends on his/her purpose of creating a new product. Similarly, with Audio options, the user can choose various functions to create his/her own products using text to speech, voice cloning, voice store and AI sound effect. Figure 2 presents all the products that Dupdub AI tool can offer people depending on his/her purposes of use.

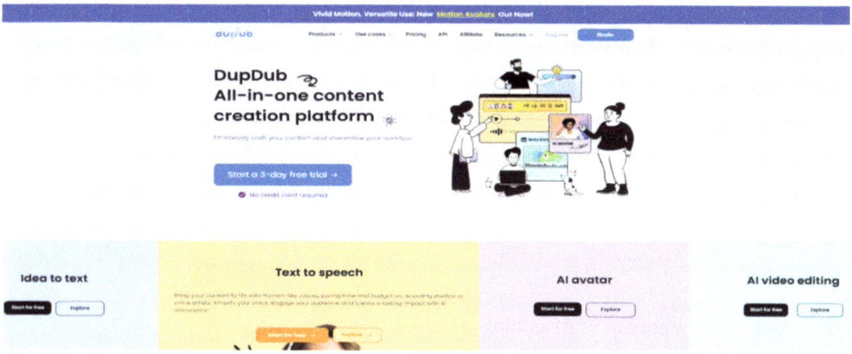


**Fig. 1** The interface of DupDub AI tool

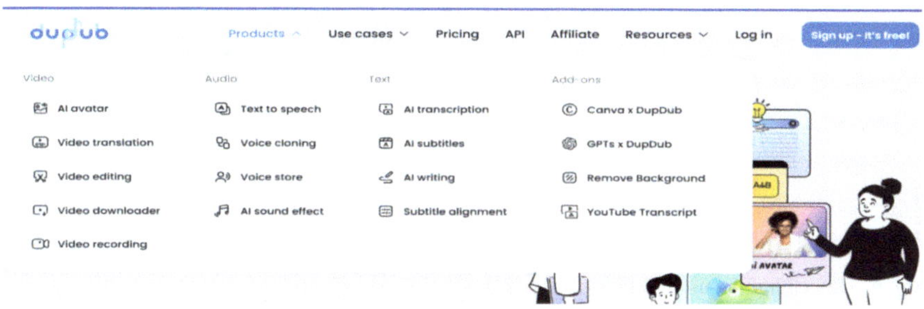


**Fig. 2** Products given by DupDub AI tool

Opening the Use case category, the user can select option Audiobook since his/her target is involved in building an English audiobook for ESL teaching. Figure 3 indicates the use cases with which Dupdub AI tool can provide the users.

There are three steps requiring users to complete. First, the user must select the purpose of using this tool by describing his/her status as a teacher/trainer, a student, a freelancer, a corporate/company, a small/medium business, or others (shown in Fig. 4).

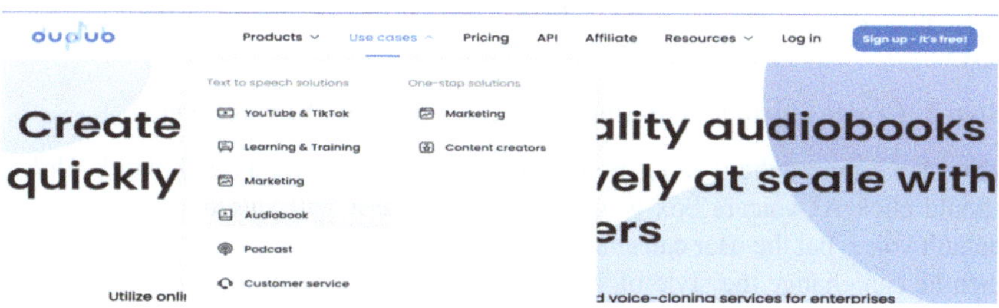


**Fig. 3** Use cases provided by Dupdub AI tool

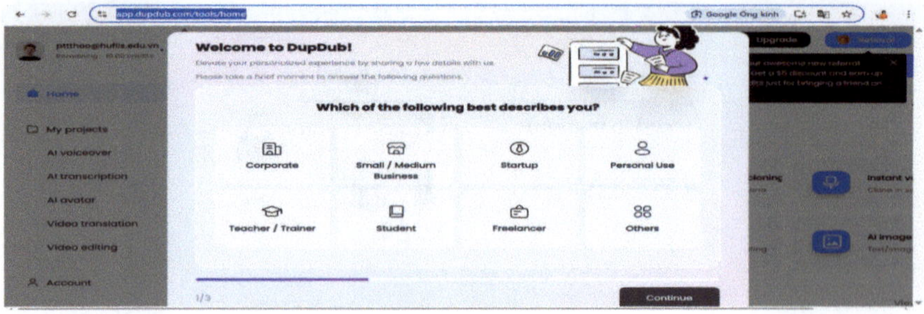


**Fig. 4** Dupdub AI's options of describing the user's status

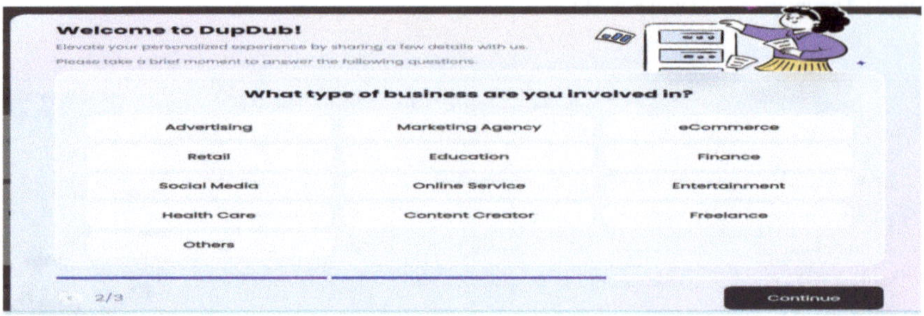


**Fig. 5** Describing the user's kinds of business

Second, if the user selects Teacher/Trainer Box, there is another table requiring the user to answer the question "What type of business are you involved in?" (seen in Fig. 5).

If the user's job related to education, he/she can select the Education box to finish the question, click Continue, and move to the next question "*What kind of content do you create or plan to create?*". Finally, he/she needs to choose the Audiobook box and submit the response to finish his/her description of using this AI tool. Figure 6 depicts the kinds of content selected by the user. The user should choose the content to submit. There are many choices including promotional ads, products introductions, social media content, learning contents, school projecct, content localization, movie clips, podcast, audiobook or others. In this case, he/she should select the Audiobook box.

**Step 2: Create a project of building audiobooks with AI voiceovers**

This step requires the user to prepare a text before creating an audiobook. He/she should click AI voicers box to get started. Maybe it will automatically choose a default voice, but the user can change the voice by clicking on the speaker icon here then he can change the style of voice, single voicer or multi-voicer (there are 754 voices for choice), gender, age and even the language and accent (see Fig. 7).

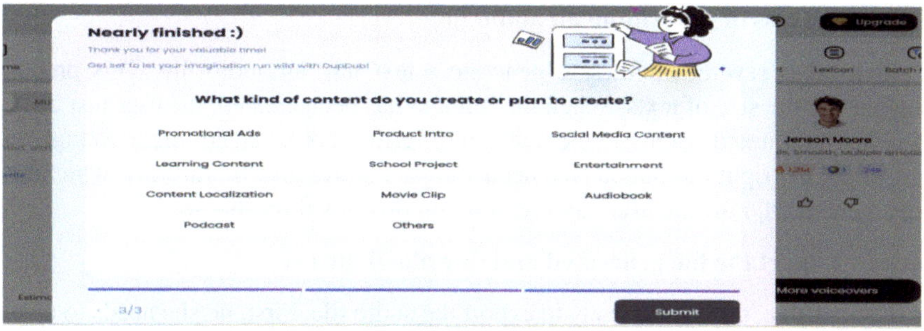


Fig. 6 Identifying the kind of creation content

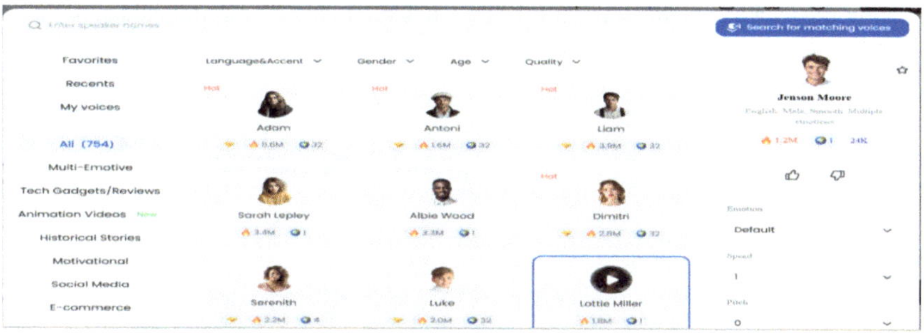


Fig. 7 Selection of voice, gender, age and language/accent

## Step 3: Upload a text/a file to generate an audiobook

To continue Step 2, after selecting the voice style, language/accent of the speakers, their age and gender, the user needs to upload a text or a file which has been prepared previously. Then he/she clicks the button Play to generate the audio file. Figure 8 shows the process of uploading a text/a file in Dupdub to start generating.

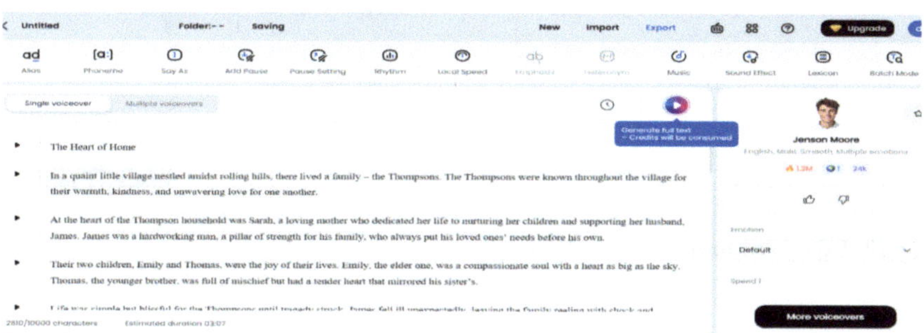


Fig. 8 Process of uploading a text/a file to generate

## Step 4: Process of generating an audio file

It often takes several minutes to generate a text into an audio file. This process depends on the size of texts. When the full text has been created, the user just clicks the Play button and listens to the audio file; furthermore, he/she might change the text to get the right file that he/she needs. Figure 9 presents the full text, which has been generated, now the user can preview and download for its use.

## Step 5: Export the file generated and download for use

In the final step, the user is ready to export the audio file. First, he/she needs to name it at the top as, for example, "The Heart at Home" story which has been automatically saved, then click Export. There are many options to export such as wave MP3, or even MP4 with subtitles, or a DOT SRT file (see Fig. 10).

Now, the user can do the same for the rest of texts till he/she has finished generating an audio output for the whole book.

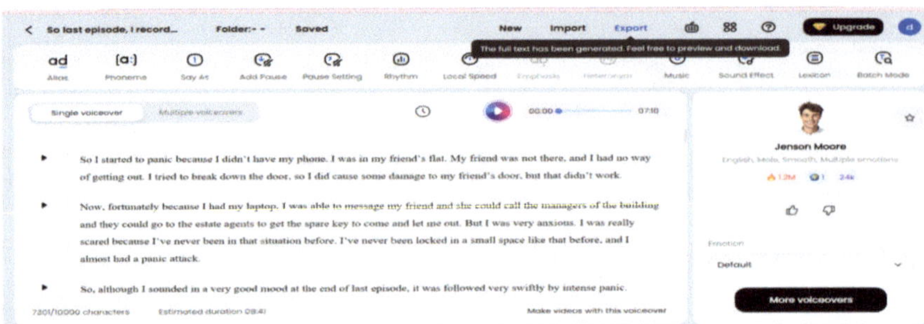


Fig. 9 Process of generating an audio text

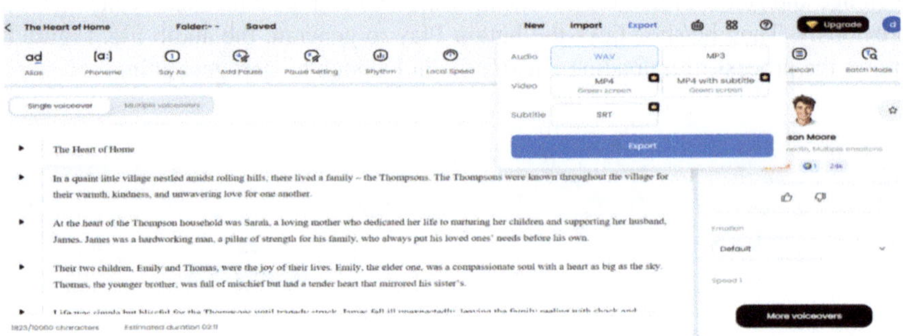


Fig. 10 Export and download the audio file

# 5 Experiment of Audiobook Application into ESL Classes and Learners' Feedback

Using audiobooks in ESL classroom was taken into great consideration at many institutions in Vietnam since an increasing number of studies on AI impacts on educational systems in general, and ESL teaching and learning in particular, have been conducted. The question of whether AI has completely supported ESL learning and teaching or not has been an endless debate. This study focuses on the use of audiobook created by AI tools in warm-up listening activities at English listening practice classes in a university of Vietnam. The research questions are posed as follows: (1) How can audiobooks be used in warm-up listening activities to enhance the listening skills of first-year English students? (2) What is students' feedback regarding the use of audiobooks in warm-up listening activities?

## 5.1 *Participants*

The participants were 155 first-year students aged from 18 to 20, who have learnt English 3–6 years at lower and higher secondary schools. These participants at English pre-intermediate level have voluntarily joined in our project to help researchers implement an experiment with AI-generated audiobooks at English listening skill practice classes. They are divided into 2 groups: (1) Experimental group including 80 students in two classes and (2) Control group with 75 students in two classes.

## 5.2 *Learning Materials*

In this experimental study, six audio files in one audiobook generated by Dupdub AI tool have been used in the warm-up listening activities at English listening skill practice classes. Table 1 presents the lessons, audio materials' description, supplementary exercises and outcomes at listening skill practice classes for 6 weeks.

## 5.3 *Data Collection and Analysis*

As mentioned above, the experiment was conducted during one semester within the Listening 1 course for 155 first-year students at HUFLIS involving four classes of two main groups (Experimental group of 80 students and Control group with 75 students). The same pre-course listening tests were given to both the experimental and control group in the first week.

**Table 1** Lesson plans with the use of audiobooks generated by AI tool for 10 min warm-up listening activities

| Lessons | Audiobook (about 5 m) | Supplementary exercises (5 min) | Outcomes | Benefits |
|---|---|---|---|---|
| Week 2<br>Objectives<br>• To expose students to authentic, natural-sounding English through audiobooks<br>• To build prediction, gist-listening, and vocabulary recall skills | Audio "*Running*" | **Vocabulary hunt**<br>Teacher gives 5 key words with their definitions or images describing these words and students listen to the audio to recognize these words | Students are able to recognize vocabulary after listening to the audio | Students are encouraged to engage with authentic material—audiobook, improve their word recognition, connect sound to meaning |
| Week 3<br>Objectives<br>• To practice listening for identifying the main idea of a story<br>• To link predictions with actual content | Audio "*I ate Collin Face*" | **True/false exercises**<br>• Students predict what the audio will be about<br>• Teacher gives 6 short true/false statements to check Ss' comprehension | Students can listen to the audio for the main ideas and predict the story and practice listening comprehension by selecting the T/F statements | Students are excited about audiobook's story and feel more confident in listening skills |
| Week 4<br>Objectives<br>• To practice listening for specific information<br>• To strengthen comprehension through Q&A | Audio "*The Heart of Home*" | **Question race**<br>Teacher gives 5 simple questions, so students listen to the audio to find answers for specific information | Students listen for specific information and know how to answer questions | Encourages active listening with a purpose, increases motivation |
| Week 5<br>Objectives<br>• To practice detail listening and word recognition<br>• To memorize vocabulary through listening | Audio "*The Farmer and His Horse*" | **Gap-filling exercises**<br>Students are given 6 short statements with 6 missing, they can fill in while listening | Students can identify specific words in the story from the audio and confirm understanding of the story's content | Students master their detailed listening skills and improves vocabulary memorization |

(continued)

**Table 1** (continued)

| Lessons | Audiobook (about 5 m) | Supplementary exercises (5 min) | Outcomes | Benefits |
|---|---|---|---|---|
| Week 6<br>Objectives<br>• To help students recognize vocabulary, sounds, or expressions while listening<br>• To practice listening to specific words or short phrases | Audio "*The Boy Who Cried Wolf*" | **Matching exercises**<br>• Teacher prepares 6 matching definitions/pictures<br>• Students listen to the audio and match the spoken words they hear to finish exercises | Students can recognize key vocabulary in authentic audio and confirm their comprehension by successfully matching words with their correct meaning | Students can enhance vocabulary, reduces stress and feel more confident when their memory has been improved |
| Week 7<br>Objectives:<br>• To listen carefully for detail and remember the events of story<br>• To improve other language skills (e.g. speaking) | Audio "T*he Amazing Thai Cave Rescue*" | **Spot the sequence**<br>Teacher provides scrambled events and students listen to a story and reorder<br>Retell sequence in 3 sentences | Students can listen to the audiobook and recognize order of events | Students can improve their story comprehension and note-taking skills and retelling stories (speaking skills) |

While both groups were observed and assisted by the researchers, only the experimental group had access to audiobooks and other related materials, and the control group continued traditional listening lessons without using any audio book in warm-up activities. The students in the Experimental group were required to follow lesson plans of warm-up listening activities for 10 min/each lesson (shown in Table 1) with designed listening activities with six AI generated audio files in one audiobook. Furthermore, this group was also instructed to listen to the audiobook in each lesson and complete all the exercises given by the researchers such as Vocabulary Hunt, T/F exercise, Question Race, Gap-filling exercises, Matching Exercises and Spot the Sequence. Then two mid-term tests were offered to students in both groups, and their scores were recorded to compare.

Quantitative and qualitative approaches were applied via study tools like Observation class, Questionnaires, Pre-course test and Mid-term test 1 and 2 to get data in this experimental study.

## 5.4 Results

Data were collected from Pre-course Test scores in Week 1. Researchers found out that a large number of students (74.2%) have many problems in learning listening and doing listening tests. Most of them (61.9%) did not pay much attention in the listening lessons at high school and high rate of students rarely practiced or did homework of listening skill (77.4%). Furthermore, the test scores were rather different in Experimental group (including 80 students) which have used audiobooks generated by AI tool in warm-up listening activities, but there was a little change in Control group (including 75 students) which used the traditional method in listening practice classes. Table 2 shows comparison of the different ranges of test scores collected from the same pre-course listening test, mid-term test 1 and mid-term test 2 in two groups.

Data from Table 2 indicates that for the experimental group, concerning the pre-course test results, majority of students (61.3%) got scores below 5.0, with only 1.3% reaching the highest scores (9.0–10.0). The students' rate obtaining scores below 5.0 in the first mid-term test has changed significantly (declining to 17.5%), while that of students' scores of 6.0–6.9 and 9.0–10.0 rose to 22.5% and 8.8% respectively. The percentage of students gaining the second Mid-term test scores continue this upward trend, with 15% achieving the highest (9–10), 22.5% (7.0–7.9), 25% (6.0–6.9) but only 6.3% of students remain in the lowest one. This evidence illustrates a steady improvement in listening comprehension when audiobooks were used in warm-up listening activities.

By contrast, it is evident to see that the control group have made little progress. In the pre-course test, 53.4% of students received scores under 5.0, and only 1.3% of them reached the highest score (9–10). In the first mid-term test, the rate of students obtaining the lowest score (<5) was still high accounting for 46.7%, but the percentage

**Table 2** Comparison of listening test scores of experiment group and control group

| Groups | Listening test scoring range | <0.5 | 5.0–5.9 | 6.0–6.9 | 7.0–7.9 | 8.0–8.9 | 9.0–10.0 |
|---|---|---|---|---|---|---|---|
| Experiment group (N = 80) | Pre-course test | 61.3 | 12.5 | 10 | 8.6 | 6.3 | 1.3 |
| | Mid-term test 1 (%) | 17.5 | 25 | 22.5 | 13.8 | 12.5 | 8.8 |
| | Mid-term test 2 (%) | 6.3 | 10 | 25 | 22.5 | 21.2 | 15 |
| Control group (N = 75) | Pre-course test | 53.4 | 17.3 | 12 | 8 | 8 | 1.3 |
| | Mid-term test 1 (%) | 46.7 | 14.7 | 13.3 | 12 | 10.6 | 2.7 |
| | Mid-term test 2 (%) | 40 | 13.3 | 13.3 | 14.7 | 13.3 | 5.4 |

of students making some progress and achieving the highest scores was small, with 2.7% (between 9.0 and 10.0). At the same time, there was not any improvement in the score obtained by students in the second Mid-term test as 40% of students got a score below 5.

These findings affirm that the use of audiobooks in the warm-up listening activities had some benefits in the listening skill development of first-year students at HUFLIS.

## 5.5 *Students' Feedback on Using Audiobooks*

After the second Mid-term Listening Test was taken, the experimental group continued to complete a questionnaire on the students' feedback on using audiobooks in listening skill practice classes which were designed with ten questions. Table 3 describes the students' feedback regarding the use of audiobooks in listening practice classes.

Data collected from the questionnaires also shows that most of students have positive attitudes toward the use of audiobooks in English listening skill practice. A very larger number of students accounting for 75.1% agreed and strongly agreed that they enjoyed listening to audiobook in practicing their English-speaking skills, and 82.5% of them agreed and strongly agreed with the statement "*I find it easy to use audiobooks for English listening practice*". This proves that students are really interested in learning English with audiobooks as they can enhance their listening comprehension in an easy way. Similarly, 68.8% of respondents confirmed that "*Audiobooks make English listening practice more interesting than traditional methods*", and 71.3% of them agreed and strongly agreed that using audiobooks helps them feel more confident in understanding spoken English. Interestingly, 23.8% of participants strongly agreed and 53.8% agreed with the benefits of improving pronunciation, a high rate (77.6%) agreed and strongly agreed with the improvement of vocabulary thanks to audiobook use. Thus, audiobooks have been highly appreciated as effective tools for ESL development.

Nevertheless, there remained some challenges of using audiobooks in listening practice lessons. About one-third of the students (31.3%) admitted that they have some difficulties in following the speed of narration (Statement 7). Most students disagreed that audiobooks are less beneficial than videos or podcasts (68.8%, Statement 9). Finally, while 21.3% had no ideas, almost half of the students (57.5%) disagreed that subtitles in audiobooks were necessary (Statement 10) for them to learn listening skills.

Overall, the findings indicate some benefits of audiobook use in ESL learning, especially in listening skill practice. For first-year students, who regarded audiobooks as good tools in developing their listening skills, audiobooks generated by AI tools can provide them with more opportunities to their ESL knowledge as well as their technological applications into real life. Therefore, in term of building and applying audiobooks in learning and teaching English, this study will provide some practical tasks in the next section.

**Table 3** Student's attitudes towards using audiobooks

| No. | Statement | Strongly agree (%) | Agree (%) | Neutral (%) | Disagree (%) | Strongly disagree (%) |
|---|---|---|---|---|---|---|
| 1 | I enjoy listening to audiobooks to practice my English listening skills | 18.8 | 56.3 | 16.3 | 6.3 | 2.5 |
| 2 | I find it easy to use audiobooks for English listening practice | 12.5 | 70 | 8.8 | 5 | 3.8 |
| 3 | Audiobooks make English listening practice more interesting than traditional methods | 21.3 | 47.5 | 12.5 | 10 | 8.8 |
| 4 | Using audiobooks helps me become more confident in understanding spoken English | 27.5 | 43.8 | 16.3 | 7.5 | 5 |
| 5 | I believe audiobooks can effectively improve my pronunciation and accent in English | 23.8 | 58.8 | 10 | 8.8 | 3.8 |
| 6 | Audiobooks provide me with a wide range of vocabulary and expressions useful for my English learning | 18.8 | 58.8 | 12.5 | 6.3 | 3.8 |

(continued)

**Table 3** (continued)

| No. | Statement | Strongly agree (%) | Agree (%) | Neutral (%) | Disagree (%) | Strongly disagree (%) |
|---|---|---|---|---|---|---|
| 7 | I often find it difficult to follow the narration speed in audiobooks | 11.3 | 20 | 18.8 | 31.3 | 18.8 |
| 8 | I lose focus easily when listening to audiobooks for English learning | 7.5 | 16.3 | 13.8 | 36.3 | 26.3 |
| 9 | Audiobooks are less useful than videos or podcasts | 7.5 | 8.8 | 15 | 45 | 23.8 |
| 10 | Without written text or subtitles, audiobooks are not effective for me in learning listening skills | 11.3 | 10 | 21.3 | 47.5 | 10 |

## 6 Practical Tasks of Creating Audiobooks with AI Tools for ESL Learners

Due to the benefits of audiobook use in listening practice classes mentioned above, it is essential to build more authentic materials to improve ESL learning. This chapter introduces some practical tasks of creating audiobooks with AI tools for ESL learners as follows.

**Task 1. Work in pairs**. Students are required to familiarize themselves with AI tools for their selection of voices and preparation of texts with different topics to create a sample of audiobook files.

**Instructions**

- Students are required to choose a short text including 150–200 words with various topics such as school, nature, environment, weather, friendship, technology and so on.
- Students should select one AI tool in order to paste or upload the text chosen.

- Students also need to choose an AI voice and adjust settings like speed, pitch and emotion of the characters in the text above.

**Task 2. Work individually**. Students should improve their listening skills with audiobooks generated by AI tools. They are required to design an audiobook used as English listening resources.

**Instructions**

- Select a short reading text on various topics as in Task 1 based on the genre of stories, articles, or dialogues.
- Use an AI tool (e.g., ElevenLabs, Dupdub, Speechify, PlayHT, WellSaid Labs, Murf AI, etc.,) to generate an audiobook.
- Practice listening to the audio to improve their listening skills.

**Task 3. Work in 3–4 student groups**. Students need to develop their technology knowledge by proofreading and building audiobooks created by AI tools. They can adjust to an audiobook available.

**Instructions**

- Students have to build a short audiobook with an AI tool (listed in Task 2).
- They need to focus on pronunciation, intonation and stressed words or phrased read by AI tool.
- Students are able to adjust the audio with punctuation or phonetic spelling.
- Present and share the output with other groups.

**Task 4. Work individually**. Students could improve their speaking skills using audiobooks to tell a story. They are required to create a short audiobook written and recorded themselves with their own voice to develop their oral storytelling.

**Instructions**

- First, students are required to write a short story with 200–300 words.
- Then they need to select an AI tool provided in Task 2 to create themselves the narration with their own voices.
- Finally, students need to add background music or sound effects if they would like to design an attractive story.

**Task 5. Work in 3–4 student groups**. Students need to improve their four English language skills including Listening, Speaking, Reading and Writing by using audiobooks created by AI tools. They are required to create a full audiobook chapter integrating all language skills.

**Instructions**

- Students are required to select a longer text including 800–1,200 words.
- They need to choose an AI tool given in Task 2 to create the narration with different voice styles, pacing and emotion of the characters in the text.
- Add optional background music or sound effects to create an impressive product.
- Publish and share the completed audiobook chapter with other students in class.

## 7 Conclusion and Pedagogical Implications

This book chapter overviews some issues relating to AI's role in creating digital resources, audiobook use with some benefits and challenges, and some main steps of building audiobooks for ESL learners to improve their listening skills. An experiment conducted at HUFLIS, Vietnam presents the application of audiobooks into warm-up listening activities in listening skill practice classes which has made some positive effects on ESL learners' improvement of language skills, learning motivation and comfort in classroom. The outcomes of this experiment affirm that students have positive attitudes towards audiobooks, which enables us to offer some pedagogical implications as follows.

- For students, they should be encouraged to use more digital learning resources such as audiobooks and eBooks in ESL learning which can promote their study motivation and self-study ability to improve their language knowledge and skills.
- For teachers, this study can be used as an example to design other digital materials, especially audiobooks to enrich the ESL teaching resources. Since the diversity of teaching authentic materials is considered one of crucial factors to create attractive, impressive and lively lessons, which can help ESL teachers enhance their teaching methods combining advanced technology knowledge with language competence. Hence, some practical tasks of creating audiobooks should be presented in the ESL course's unit description to renovate the teaching strategies in digital education.
- For university, thanks to recent advancement in the 4.0 technologies leading to a rapid growth of both availability of audiobooks and an increasing number of audience, it is necessary to release some policies of education development, among which the integration of audiobooks into ESL learning/teaching, designing more authentic resources with audiobook usage should be paid more attention in the digital era. Hence, both learners and teachers could improve not only their language skills, but also their technology knowledge which is of necessity to meet the increasing educational and social needs nowadays.

**Acknowledgements** We used artificial to proofread this manuscript.

**Ethics** This manuscript has not been published anywhere or is under consideration for publication.

## References

Batista, J., Mesquita, A, & Carnaz, G. (2024). Generative AI and higher education: Trends, challenges, and future directions from a systematic literature review. *Information, 15*(11).

Chang, A. C. S. (2011). The effects of reading while listening to audiobooks: Listening fluency and vocabulary gain. *Asian Journal of English Language Teaching, 21*, 43–64.

Chang, A. C. S., & Millett, S. (2015). Improving reading rates and comprehension through audio-assisted extensive reading for foreign language learners. *Reading in a Foreign Language, 27*(2), 263–283.

Chen, Y. (2024). Enhancing language acquisition: The role of AI in facilitating effective language learning. In *Proceedings of the 2024 3rd international conference on humanities, Wisdom education and service management* (HWESM 2024) (pp. 593–600).
Farrelly, T., & Baker, N. (2023). Generative artificial intelligence: Implications and considerations for higher education practice. *Education Sciences, 13*(11), 1109. https://doi.org/10.3390/educsci13111109
Ferris, D., & Tagg, T. (1996). Academic listening/speaking tasks for ESL students: Problems, suggestions, and implications. *TESOL Quarterly, 30*, 297–320. https://doi.org/10.2307/3588145
Fjelland, R. (2020). Why will general artificial intelligence not be realized. *Humanities and Social Sciences Communications*, *7*(10). https://doi.org/10.1057/s41599-020-0494-4
Geraldin, B., Dela, C., & Jaya, R. (2022). The effects of AI in various spheres of life. *Technoarete Transactions on Advances in Computer Applications, 1*(2).
Ghafar, Z. N., Salh, H. F., Abdulrahim, M. A., Farxha, S. S., Arf, S. F., & Rahim, R. I. (2023). The role of artificial intelligence technology on English language learning: A literature review. *Canadian Journal of Language and Literature Studies, 3*(2), 17–31. https://doi.org/10.53103/cjlls.v3i2.87
Gilakjani, P. A., & Sabouri, N. B. (2016). Learners' listening comprehension difficulties in English language learning: A literature review. *English Language Teaching, 9*(6), 123–133.
Hamouda, A. (2013). An investigation of listening comprehension problems encountered by Saudi students in the EL listening classroom. *International Journal of Academic Research in Progressive Education and Development, 2*(2), 113–115.
Handayan, M., Luciana, & Sugiharto, S. (2025). Language materiality and transmodal production in EFL classroom: The case of Dayaknese students. *Jurnal Arbitrer*, *12*(3), 322–355. https://doi.org/10.25077/ar.12.3.322-335.2025
Harris, D. D. (2001). An inquiry into the influence of taped books on the oral reading performance of struggling third-grade readers. ProQuest Dissertations and Theses, Louisiana State University and Agricultural and Mechanical College. http://search.proquest.com/docview/304700784?accountid=16935
Jacobs, T. (2006). Teaching with audiobooks. Article Alley. www.articlealley.com
Javad, K. H., & Leila, A. A. (2015). Attitudes toward using the Internet for language learning: A case of Iranian English teachers and learners. *International Journal of Research Studies in Educational Technology*, *4*(1), 63–78. https://www.researchgate.net/publication/272399631_Attitudes_toward_using_the_Internet_for_language_learning_A_case_of_Iranian_English_teachers_and_learners
Kartal, G., & Simek, H. (2017). The effects of audiobooks on EFL students' listening comprehension the reading matrix. *An International Online Journal*, *17*(1).
Khalilova, L., Allayorov, A., & Yusupov, A. (2025). The benefits of AI in education: Revolutionizing teaching. *Академические Исследования в Современной Науке, 4*(6), 36–41.
Koskinen, P. S., Blum, I. H., Bisson, S. A., Phillips, S. M., Creamer, T. S., & Baker, T. K. (2000). Book access, shared reading, and audio models: The effects of supporting the literacy learning of linguistically diverse students in school and at home. *Journal of Educational Psychology, 92*(1), 23–36.
Kothai, N. (2024). AI powered tools for academic research. In M. Deivam, J. Neeraj, & P. N. Manoj (Eds.), *Artificial intelligence in education.* Mithra Dev Publication.
Krylov, E. A. (2023). Language learning 2.0: The role of artificial intelligence in facilitating second language acquisition, *4*(1), 5–12.
Mukhallafi, T. R. A. (2020). Using artificial intelligence for developing English language teaching/learning: An analytical study from university students perspective. *International Journal of English Linguistics, 10*(6), 40. https://doi.org/10.5539/ijel.v10n6p40
Qian, H., Chunlan, L., Li, L., & Shuang, T. (2025). Evaluating the quality of AI-generated digital educational resources for university teaching and learning. *Systems*, *13*(3), 174. https://doi.org/10.3390/systems13030174

Renandya, W. A., & Farrell, T. S. C. (2012). The Tape is too fast. Modern English teacher. *ELT Journal, 65*(1), 52–59.

Rizal, D., Masruroh, S., Rahman Syah, R. F., Fathina, I. F., & Zakariyah, S. (2022). Audiobooks as media to increase listening and speaking skills: A qualitative systematic review. *ETERNAL (English Teaching Learning and Research Journal), 8*(1), 200–216. https://doi.org/10.24252/Eternal.V81.2022.A14

Saka, Z. (2015). The effectiveness of audiobooks on pronunciation skills of EFL learners at different proficiency levels. Unpublished MA thesis, Bilkent University.

Saraswaty, D. R. (2018). Learners' difficulties & strategies in listening comprehension. *English Community Journal*, 2(1), 139–152, 139. http://jurnal.um-palembang.ac.id/englishcommunity/index

Serafini, F. (2004). *Audiobooks and literacy: An educator's guide to utilizing audiobooks in the classroom.* Random House Inc.

Sharma, S. (2025). *How to make an audiobook with AI?* https://murf.ai/blog/create-a-magical-audiobook-experience-with-ai-voices

Stone-Harris, S. (2008). The benefit of utilizing audiobooks with students who are struggling readers. Education. ProQuest dissertations and theses. http://search.proquest.com/docview/304389547?accountid=16935

Tagninezhad, A., Khalifah, A., Nabizadeh, A., & Shahab, S. (2015). Examining the influence of using audiobooks on the improvement of sound recognition and sound production of Iranian EFL learners. *International Journal of Languages' Education and Teaching, 3*(3), 28–42.

Tran, T. Y., Pham, T. K., & Tran, M. N. (2021). *Use of audiobooks to improve listening comprehension for English majors at Thai Nguyen University of Education.* https://doi.org/10.29322/IJSRP.11.06.2021.p11480

Wolfson, G. (2008). Using audiobooks to meet the needs of adolescent readers. *American Secondary Education, 36*(2), 105–114.

Woodall, B. R. (2010). Simultaneous listening and reading in ESL: Helping second language learners read (and enjoy reading) more efficiently. *TESOL Journal, 1*(2), 186–205.

Zanyar, N. G., et al. (2023). The role of artificial intelligence technology on English language learning: A literature review. *Canadian Journal of Language and Literature Studies, 3*(2), 17–31.

**Phan Thi Thanh Thao** holds a Ph.D. in Natural Language Processing and is currently a senior lecturer at the Faculty of English, University of Foreign Languages and International Studies, Hue University (HUFLIS), Vietnam. Her research interests lie in the science of language, especially computational linguistics, machine translation, computer-assisted translation (CAT) tools, translation studies, and corpus linguistics. In addition, her academic background is closely related to TESOL, with a strong emphasis on the integration of technology, particularly Computer-Assisted Language Learning (CALL).

# AI-Mediated Learning and Interaction

# Chatbot-Human Interaction in Vietnamese EFL Speaking Classes: Exploring the Unchartered Territory

**Ho Thi Hien, Nguyen Van Huy, and Nguyen Thi Bich Hanh**

**Abstract** AI voice-based chatbots (AVCs) are able to act as promising learning assistants, enhancing speaking proficiency as conversation partners. However, little is known about how learners and AVCs actually interact with each other in such conversations. Drawing on (Long, Handbook of second language acquisition, Academic Press, 1996) Interaction Hypothesis, this study investigates how Vietnamese university students interact orally with ChatGPT, an AVC, in speaking classes. Through an exploratory, mixed-methods research design using both reflective surveys and screen recordings, the chapter examines the types of linguistic adjustments ChatGPT made during chatbot-human interactions, the ways both the students and ChatGPT negotiated for meaning, and how the students modified their language outputs during conversations with ChatGPT. The results reveal that, overall, students have a positive attitude towards integrating ChatGPT in practicing speaking skills. The results also show that both students and ChatGPT have adjusted their language to negotiate for meaning, such as adjusting speech rate, asking for clarification and repetition, using paraphrases, and elaboration. These results highlight the engagement of chatbot-human interactions and their benefits for students' speaking practice and performance. The chapter concludes by discussing the implications of these findings for using AVCs as an effective conversation partner to boost students' speaking skills and suggests pathways for further research.

**Keywords** AI voice chatbot · ChatGPT · Vietnam · EFL speaking · Interaction hypothesis · Chatbot–human interaction

H. T. Hien · N. Van Huy (✉)
Hue University of Foreign Languages and International Studies, Hue, Vietnam
e-mail: nguyenvanhuy@hueuni.edu.vn

H. T. Hien
Industrial University of Ho Chi Minh City, Ho Chi Minh City, Vietnam

N. T. B. Hanh
Ha Tinh University, Hà Tĩnh, Vietnam

 
H. P. Bui and L. P. H. Huong (eds.), *Artificial Intelligence in Education and Academic Research*, Studies in Computational Intelligence 1269,
https://doi.org/10.1007/978-3-032-18392-7_6

## 1 Introduction

Speaking is one of the most complex skills for language learners to master (Brown, 2001). In conventional English classes, learners practice speaking skills by engaging in conversational interaction with an interlocutor (i.e., a peer or a teacher). Such a teaching context, on the one hand, fosters learners' speaking skills through learners' effort to use the target language to make themselves understood, or to understand what the interlocutor means; it often poses limitations. For example, students do not have sufficient opportunities to practice speaking meaningfully either with peers or teachers inside the classroom (Al-Tamimi et al., 2020; Gan, 2012; Hwang et al., 2024). Other factors, such as large class sizes, limited teacher attention, and a lack of authentic communication situations, also limit the effectiveness of speaking instructions (Ur, 1996). Nunan (2003) further notes that students often experience high levels of anxiety and low confidence when speaking, especially in traditional teacher-led speaking classes.

In Vietnam, teaching speaking skills also faces numerous challenges, resulting in student outcomes often falling below expectations, despite national commitments to foreign language proficiency (Hayden & Le Tran, 2024). Firstly, EFL learners often lack sufficient opportunities to practice speaking within the classroom, particularly students from regional areas who may have limited exposure to spoken English (Nguyen & Burke, 2025). Secondly, students often feel scared of making mistakes and being judged by other people. Thus, they prefer using Vietnamese in speaking lessons, even though students with lower English proficiency may be inattentive or unmotivated (Le & Thi, 2022). Furthermore, Vietnamese-speaking teaching practice also faces similar practical issues like the aforementioned context, such as large class sizes, noisy classrooms, and limited access to authentic teaching materials that further hinder teaching and learning speaking progress (McAleavy et al., 2018; Nguyen & Burke, 2025; Pham et al., 2023).

Recently, technological advances have assisted teachers in increasing the effectiveness of teaching speaking skills. AI voice-based chatbots (hereafter, AVCs) are tools that combine artificial intelligence and natural language processing to facilitate speech-based interactions that mimic human conversations (Koç & Savaş, 2025; Pérez et al., 2020). Although research on the integration of AVCs in speaking instructions is still in its infancy (Du & Daniel, 2024), the findings primarily support that common AVCs such as Replika, ChatGPT, Andy, and Mondly can be promising learning assistants, enhancing speaking proficiency (Azizimajd, 2023; Çakmak, 2022). AVCs are designed to function as conversation partners engaging in dialogues with users and interpreting user inputs using natural language to provide responses and feedback accordingly (Adamopoulou & Moussiades, 2020).

To address such problems, researchers have explored the potential of using different technological tools and applications in developing learners' speaking skills. Some studies have shown that using automatic speech recognition (ASR) applications such as ELSA Speak, Duolingo, ILI, and NOVO can benefit students' speaking skills in different ways. For example, these applications can provide instant pronunciation feedback, help learners notice phonological errors, and foster autonomous, self-paced speaking practice, leading to learners' improvement in pronunciation accuracy, fluency, and oral confidence (Bashori et al., 2024; Pham & Pham, 2025; Sun, 2023). However, using ASR applications for practicing pronunciation and speaking skills posed limitations. For example, most of the ASR systems are trained to read monological speech rather than authentic interaction and negotiation as a natural conversation (Kalimullah & Tabassum, 2025; Liu et al., 2025).

Other studies have examined the positive impacts of AI voice-based chatbots on students' speaking skills. Learners perceived that AVCs such as Replika, Andy, ChatGPT, and Mondly helped them feel more comfortable, motivated, and confident practicing speaking without the pressure of being judged (Ding & Yusof, 2025; Duong & Suppasetseree, 2024), and increased willingness to communicate (Fathi et al., 2024; Yang et al., 2022). AVCs also have positive impacts on learners' improvement of pronunciation, fluency, and vocabulary with personalized and timely feedback (Ding & Yusof, 2025; Duong & Suppasetseree, 2024; Yang et al., 2022). Nevertheless, learners also reported some potential issues that impede the optimal effectiveness of interacting with AVCs, such as feeling anxious and frequently demotivated due to communication failures or unnatural language content from chatbots (Çakmak, 2022; Duong & Suppasetseree, 2024; Jeon, 2024). However, as oral interaction is a complex process in which learners need to employ different techniques (i.e., clarification, confirmation, and meaning negotiation) to accomplish communicative goals (Long, 1996), there seems to be a lacking focus on conversational/interactional mechanism in the previous studies, which might provide more insights into why learners encounter these challenges during their interaction with chatbots and what can be done about them. Therefore, there is a need for a better understanding of how learners and chatbots interact as conversation partners. In addition, there are quite limited studies on the use of AVCs in English classes within the Vietnamese educational context. To the best of the researchers' knowledge, there are currently two reported studies on integrating AVCs in teaching speaking skills. The first one was conducted with 38 intermediate-level EFL learners at a private university (Nguyen, 2025), and another study investigated the effects of using AVCs on undergraduates' speaking skills (Duong & Suppasetseree, 2024). However, these studies focused on the impacts of interacting with AVCs on learners' speaking skills rather than exploring how learners interacted with the AVCs.

Our study aimed to fill this gap by exploring how Vietnamese university students interact orally with an AVC, specifically ChatGPT, to practice speaking skills in class. We designed our study following the Interaction Hypothesis (IH) developed by Long (1996), an American linguist who made influential contributions to second language acquisition theory and practice. This theory emphasizes that second language acquisition is promoted effectively through meaningful interaction. In such communicative contexts, learners need to ask for clarification, confirmation checks, repetition checks (i.e. negotiation of meaning) which leads to the output modification. We addressed three research questions

1. What types of linguistic adjustments does ChatGPT use to modify its input during the chatbot–human interaction?
2. How do Vietnamese university students and ChatGPT negotiate for meaning during their conversation with ChatGPT?
3. How do Vietnamese university students modify their language output during their conversation with ChatGPT?

## 2 Literature Review

### 2.1 AI Voice-Based Chatbots

Simply put, an AI voice chatbot is an intelligent system that uses artificial and large language management to converse orally with learners, engaging them in real-time spoken interaction. This can be used to develop learners' English communication skills. Koç and Savaş (2025) define AVCs as artificial intelligence systems that are capable of emulating conversational humans through text- or speech-based interactions. They use Natural Language Processing to understand and respond to users' questions and then facilitate two-way communication orally. They are virtual agents that can understand and answer questions from users all day and support a large number of users at the same time (e.g. Adamopoulou & Moussiades, 2020; Belda-Medina & Calvo-Ferrer, 2022). Çakmak (2022) also proposes that AVCs can communicate with students as native speakers, which enhances students' interest in speaking. Among many newly-invented AVCs, ChatGPT's voice mode has been adopted for speaking practice in language classrooms in different education contexts (Lim, 2024; Muniandy & Selvanathan, 2025; Zou et al., 2025) and has proven to be beneficial for language learners in speaking practice.

### 2.2 ChatGPT and Its Voice Mode

ChatGPT is a powerful natural language processing system designed to generate human-like conversations by understanding their context. This conversational format enables ChatGPT to answer follow-up questions, provide feedback, detect errors, and reject inappropriate requests. ChatGPT was first launched in November 2022 by OpenAI, a company specializing in AI research and development whose mission is to ensure the benefits of artificial general intelligence for humanity. ChatGPT was built on the Generative Pre-trained Transformer architecture and trained on massive text datasets that enable it to process and generate text that is similar to human-written text (Pavlik, 2023). Initially, ChatGPT was powered by the GPT-3.5 model; it then quickly became significantly attractive for its ability to generate coherent and contextually appropriate text-based responses. The latest version is ChatGPT-5 which was released by OpenAI on August 7, 2025. This version is far more functional than its original one for its capability of improved reasoning, contextual understanding, and multimodal processing (e.g., text, audio, and images) (OpenAI, 2025b).

In September 2023, the voice-based feature of ChatGPT was rolled out for paid users on mobile phones, which marks a significant milestone in this conversational agent's development. The incorporation of ChatGPT's speech function facilitates two-way audio interaction; thus, users can interactively engage in spoken dialogues with a virtual agent (OpenAI, 2025a). For example, ChatGPT's voice-based feature can give real-time feedback on pronunciation, assist language learners in practicing listening comprehension, and improve speaking fluency by engaging in oral conversations with exposure to diverse accents. Adaptive feedback further enhances the learning experience. Previously, the voice-based function was only accessible on mobile and desktop apps; however, starting in 2024, OpenAI released notes of updates regarding the availability of the advanced voice feature for ChatGPT on the website. At the present time, ChatGPT's voice-based feature is now available in all types of devices for both free and paid version, although there are certain differences in their capacity (e.g. the free accounts only offer 5 minutes for voice chat while the subscription ones offer almost unlimited voice chat time; or the paid version include nine different voice modes of both female and male with different tones and moods for users to choose, whereas, the free accounts only offer standard voice).

## 2.3 *The Interaction Hypothesis*

The Interaction Hypothesis (IH) was developed by Long (1996), an American linguist who made influential contributions to second language acquisition theory and practice. The IH emphasizes the importance of three constructs in SLA: comprehensible input, negotiation of meaning, and modified output.

### 2.3.1 Comprehensible Input

Krashen (1982) defines comprehensible input as "$i + 1$". It refers to the language produced by the more competent speaker in the dialogue, which is slightly above the learners' current level of proficiency. Still, it can be made understandable with context support. In conventional communication, input often comes from a more capable human interlocutor, such as a teacher or a native speaker, who adapts their language to make it comprehensible for the learners. According to Long (1996), there are two primary ways that the more competent speaker can modify input to make it understandable for learners, including simplification and elaboration. *Simplification* is the process of making input easier to understand by reducing linguistic complexity, such as shortening sentences, using more popular vocabulary, and speaking more slowly and clearly. *Elaboration* aims at making the input comprehensible by adding information such as paraphrasing, providing examples, and using gestures and nonverbal cues. With the integration of AVCs, our study examined how the chatbot adapts its language to make it more understandable for learners as a conversation partner.

### 2.3.2 Negotiation of Meaning

Long (1996) defines negotiation of meaning as a process in which learners and fluent speakers work together to understand each other by listening, checking what the other person means, and then adjusting their words, sentences, or ideas until they both understand clearly the message. He suggests that negotiation of meaning is the key process through which conservationists can either avoid conversational trouble or repair communication breakdowns that may occur. Common meaning-negotiation strategies include repetitions, comprehension checks, confirmation checks, and clarification requests. *Repetition requests* refer to questions that the listener wants the speaker to say the whole or a part of the utterance that was just repeated, so that it can be better understood. *Clarification requests* are questions that listeners use to signal they have not understood certain pieces of information and are asking for elaboration or reformulation, such as an example, explicit explanation, or definition of new words. *Confirmation checks* refer to a sort of question or a move that speaker uses to confirm what they heard is correct. Lastly, *comprehension checks* refer to techniques that speakers use to ensure that the listener understands the message correctly. Negotiation of meaning leads to interactional adjustment by the more competent interlocutors, facilitating acquisition by effectively linking input, internal learner capacities, and output. Traditionally, these linguistic adjustments are provided by native speakers, teachers, or a more capable peer; however, in the context of this study, the conversation partner was an AVC.

### 2.3.3 Modified Output

Modified output occurs when learners change or improve their utterance after realizing that what they initially said was unclear or incorrect. Long (1996) explains that when communication breaks down, learners often adjust their language to make it more accurate or understandable. Modifying the output is vital in second language learning because it gives learners the chance to test their language, notice their mistakes, and try new forms (Swain, 1985). Pica et al. (1989), as cited in Foster's (1998), categorized learners' output modifications into four types with specific operationalized descriptions. First, learners' output can be modified by changing the meaning of what they say (i.e., *semantic modification*). For example, learners may use synonyms, paraphrase, or provide examples to clarify what they mean. The second type of output modification is *morphological* change, which focuses on how learners correct their use of word forms. Subsequently, learners can also modify their output production by improving pronunciation (i.e., *phonological modification*). Finally, *syntactic modification* or fixing sentence structure is another way that learners can use to change their language use to make themselves understood. For example, if a learner says "He go to school yesterday," then he might self-correct it to "He went to school yesterday." Modifying output helps learners move from noticing errors to producing more accurate and fluent language.

# 3 Methodology

## *3.1 Research Design*

An exploratory mixed-method research design was employed in this study. It combined quantitative and qualitative methods to provide a comprehensive examination of chatbot–human interaction in the EFL classroom. As this topic is currently in its early stage (Du & Daniel, 2024), we selected the exploratory research to understand the nature of chatbot–human interaction before taking further steps in studying this novel area. Mixed-methods design is suitable for researching the technology-mediated language teaching area because it allows researchers to collect measurable data and insightful information into learners' experiences (Creswell & Clark, 2018).

## *3.2 Participants*

The study was conducted in a real classroom context with three intact classes at the Industrial university of Ho Chi Minh university. The contents of all lessons followed the coursebook to ensure the authenticity and alignment with the curriculum as per the institution's regulations. The research was approved by the institution, faculty, and division leads via a gatekeeper letter before it was officially launched. The students did the tasks simultaneously in the classroom. According to Mackey and Gass (2015), conducting research in intact classes guarantees ecological validity because it preserves the natural classroom setting and avoids disruptions caused by artificial arrangements. The teacher acted as a researcher to avoid the presence of a stranger in the classroom and make sure that the lessons could run as planned. To reduce the subjectivity of the researcher's dual role, a Canadian teaching assistant was invited to participate in the class to observe, and cross-checked that the speaking tasks, which were designed and conducted following the syllabus, and the class procedure were consistently organized over weeks.

There were 42 English-major sophomores in each class; thus, a total number of 126 students from the three classes (with pseudo names as classes A, B, and C). However, 12 students were absent on the day that the survey was administered. Therefore, the final number of respondents was 114, including 79 female (69%) and 35 male (31%) students. They came from various regions across the country, including the highland, countryside, and urban provinces. Their English proficiency levels, as self-reported at the beginning of the semester, ranged from lower-intermediate (A2) to upper-intermediate (B2), with the majority of them at levels A2 to B1. Regarding their experience of having conversations with chatbots, 85 students (75%) had never tried it before, while the remaining 29 students (25%) had heard about it but had not used it to practice speaking seriously.

## 3.3 *The "GPT Talk Time" Project*

To explore how EFL students interacted with AI voice-based chatbots, a 6 week project, named "*GPT Talk Time*", was conducted with the approval of the institution and faculty leads. It was designed to integrate chatbot-based speaking practice into the regular curriculum. Each lesson followed a fixed procedure and lasted for 90 min per week. First, the teacher introduced the topic, taught vocabulary and grammatical structures, and suggested ideas related to the topic with the assistance of a teaching assistant. Then, instead of talking to a human partner, students practiced speaking tasks with ChatGPT for 30 min. Finally, students shared the answers with the whole class and learned from feedback. The Plus subscription of ChatGPT was used for six weeks. The project was conducted in a lab room with computers, headphones, and an internet connection.

Every week during the GPT Talk Time project, students and ChatGPT participated in an information gap task whose content was based on the textbook New Interactions 2: Listening and Speaking (Tanka & Most, 2020) to align with course objectives. Students and ChatGPT were given different information and needed to exchange information in order to complete the task. Specifically, in each lesson, students were provided with a printed handout including task details and instructions. Another version of the handout was sent to the students' computers and uploaded to the chat box of ChatGPT. ChatGPT's handouts included the instructions, prompt, and task details for it to understand how and what to do during the conversations. Before each *GPT Talk Time* session, the teacher shared the screen to explain the handout and instructions clearly to ensure that all students understood the tasks and how to carry them out. Information gap activities have long been recognized as effective for stimulating negotiation of meaning and pushing learners to request clarification, confirmation, and elaboration (Doughty & Pica, 1986).

## 3.4 *Instruments*

### 3.4.1 Reflective Survey

To explore how students and the chatbot interacted with each other, students were required to complete a structured reflective survey. The survey consisted of two sections with a number of 17 items: comprehensible input (6 items), and modifying input (11 items). The development of items in the two sections was based on the key constructs of Long's (1996) Interaction Hypothesis. Cronbach's alpha coefficients were 0.76 for comprehensible input and 0.84 for modifying input items, demonstrating acceptable to high internal consistency.

The survey was first designed in English by the first researcher and then sent to the other two researchers to revise and reword. When the English version was finalized, it was translated into Vietnamese to facilitate easy understanding for the respondents. Before being administered to the participants, the survey was piloted with 10 students who did not participate in the project to check for clarity, time to respond, and potential issues.

### 3.4.2 Screen-Recordings

During the technical training sessions in Week 1 and Week 2 of the project, all students were informed that their sessions would be screen-recorded. They were guided to start and stop screen-recording with the OBS application, which is free and user-friendly for video recording. It can capture what is shown on the computer screen (i.e., the interface of ChatGPT) and can also record the device sound and the speaker's sound with very high-quality output. Using recordings serves several purposes, helping researchers collect live and naturalistic records of learner performance. They can be used to analyze multiple times, which can not be done with classroom observation (Mackey & Gass, 2015). In addition, audio recordings ensure that not only the spoken language but also timing, pauses, and repair sequences are captured, which are necessary for measuring fluency and discourse management of speaking skills (Foster et al., 2000). After being recorded, the recordings were automatically saved in a folder on the computer. The students were then requested to upload their recordings to a Google Drive folder created by the instructor. To ensure the privacy and confidentiality of the screen-recordings, several steps were taken before, during, and after the project. First, the participants provided informed consent for screen recording without a camera on (i.e., participants' faces and real names would not appear in the recorded videos). In addition, after being downloaded, the recordings were anonymized as class-number of the student-number of recording (e.g., A3-Recording 1; B5-Recording 2). This pseudonymization method helped us retrieve the recordings easily and organize them logically. Finally, all recordings were downloaded and securely stored in the first researcher's Google Drive folders. All the folders were password-protected and could be accessed by the assigned researchers only.

## 3.5 Data Collection

At the beginning of the semester, the teacher introduced the project and its purpose to all students in three classes. After that, technical training was provided to all students to ensure that they know how to start and stop screen-recording with the OBS Studio application. Students were also guided to take some personalization settings for a ChatGPT account, log in, and make a trial conversation with ChatGPT on a given situation. After this week, the students reported any technical issues they may have encountered and any worries about practicing speaking with a virtual partner.

In week 2, students participated in a speaking task with ChatGPT, following a step-by-step procedure provided in a printed handout. The purpose was to familiarize themselves with conducting a conversation with the chatbot. In addition, the techniques for meaning negotiation were also introduced to the students to help them solve the communication breakdown. ChatGPT Plus accounts and self-practice tasks were also assigned to students to help them become more familiar with chatbot–human interaction in the coming week. From week 3, screen recordings were collected. The reflective survey on how students interacted with the chatbot was administered to students via Google Form in week 6 of the semester. The completion of the survey took approximately 15 min.

## 3.6 Data Analysis

In order to answer research question 1, descriptive statistics (mean, standard deviation, and distribution) of responses to the survey items were calculated using SPSS version 26 to examine the perceived frequency and types of linguistic adjustment that ChatGPT made to make it comprehensible. For research questions 2 and 3, which aimed to capture the actual and live interactional moves of both students and ChatGPT to negotiate meaning, we randomly selected five recordings of five students interacting with ChatGPT in an information gap task. The recordings were transcribed by NotebookLM, and then cleaned and double-checked by the researchers.

Segments of text were coded into communication units (C-units) (Loban, 1976). A C-unit consisted of "an independent clause with its modifiers which cannot be further divided without the loss of its essential meaning" (Loban, 1976, p. 9). In other words, a C-unit is an independent clause and its associated dependent clauses. A C-unit can be segmented as a sentence or a word as long as the utterance indicates a complete thought. Loban categorizes C-units into three types: a standalone grammatical statement, a short reply to a question, or a word when given in answer to a question. Table 1 provides examples of C-unit counting.

Previous researchers such as Foster (1998), Yufrizal (2001), and Channell et al. (2018) have used C-units to identify negotiation-of-meaning utterances and output modification in interactional tasks. This approach allows researchers to classify all the meaning negotiation and modified output moves efficiently and reliably within the

**Table 1** Example of communication unit (C-unit) counting

| Utterances | Number of C-units |
|---|---|
| 1. I know a boy with red hair | 1 |
| 2. When I was fixing ready to go home my mother called me up in the house// and [I, I, have to] have to get my hair combed | 2 |
| 3. • Have you been sick? // • No | 2 |

available time (Foster, 1998). Apart from C-unit analysis, other researchers have used text unit (T-unit) (Hunt, 1966) and Analysis of Speech unit (AS-unit) (Foster et al., 2000) to analyze learners' speech. While these methods are considered to be valuable for fine-grained syntactic or Complexity, Accuracy, Fluency (CAF) analysis, it was argued that these methods could be very labor-consuming and might reduce reliability in spontaneous learner talk because they required separating every coordinated or interrupted clause (Foster et al., 2000).

To analyze the recordings and explore the incidence of negotiations for meaning, output modifications for questions 2 and 3, we drew on Long's (1996) Interaction Hypothesis to define the constructs, while the analytical procedures were adopted from Foster's (1998) measurement of C-units. First, we measured the moves of negotiation for meaning by screening the C-units and classifying them into 3 categories: (1) *clarification checks*, (2) *confirmation checks*, and (3) *comprehension checks*. For modified output measurement, C-units were grouped into four categories (1) *semantic modifications*, (2) *morphological modifications*, (3) *phonological modifications*, and (4) *syntactic modifications*. Finally, the numbers of moves were calculated manually to analyze the distribution and extractions from live conversations.

# 4 Results and Discussion

## *4.1 ChatGPT's Linguistic Adjustments to Modify Its Input*

Overall, participants reflected that the level of English that ChatGPT used was slightly difficult for them to follow (see Table 2), especially the speech rate and length. They reported that ChatGPT spoke quite fast, which is the hardest thing impeding their interaction, with over 70% of them rating it as difficult or very difficult. ChatGPT's long turn length was the second common problem, with half of the students struggling to keep up with the extended input. In contrast, vocabulary, grammar, and ideas used by ChatGPT were rated at a moderate level by the majority of the participants. Students appeared to have different experiences with ChatGPT's pronunciation; 26.3% of them found ChatGPT's pronunciation easy or very easy to understand, 42.1% moderate, and 31.6% found it hard to understand.

**Table 2** Perceived level of input comprehensibility

| Comprehensible input items | N | Very easy | Easy | Moderate | Difficult | Very difficult | M | SD |
|---|---|---|---|---|---|---|---|---|
| | | 1 | 2 | 3 | 4 | 5 | | |
| 1. ChatGPT's vocabulary use | 114 | 1<br>0.9% | 10<br>8.8% | 59<br>51.8% | 41<br>36% | 3<br>2.6% | 3.31 | 0.71 |
| 2. ChatGPT's grammar use | 114 | 0<br>0% | 11<br>9.6% | 68<br>59.6% | 31<br>27.2% | 4<br>3.5% | 3.25 | 0.67 |
| 3. ChatGPT's pronunciation | 114 | 9<br>7.9% | 21<br>18.4% | 48<br>42.1% | 30<br>26.3% | 6<br>5.3% | 3.03 | 0.99 |
| 4. ChatGPT's speech rate | 114 | 0<br>0% | 5<br>4.4% | 28<br>24.6% | 52<br>45.6% | 29<br>25.4% | 3.92 | 0.82 |
| 5. ChatGPT's turn length | 114 | 0<br>0% | 4<br>3.5% | 53<br>46.5% | 48<br>42.1% | 9<br>7.9% | 3.54 | 0.69 |
| 6. ChatGPT's ideas | 114 | 2<br>1.8% | 14<br>12.3% | 67<br>58.8% | 27<br>23.7% | 4<br>3.5% | 3.15 | 0.74 |

*Note* N: number of participants, M: mean; SD: standard deviation; italicized *digits* indicate the raw counts for each response option

This result indicated that students have different reflections on the levels of language input that ChatGPT used. While a small percentage of students found that ChatGPT's language is easy for them to understand, nearly half of the students found ChatGPT's input moderately comprehensible. However, many other students still struggled to understand ChatGPT's input. For example, 44% students found ChatGPT's vocabulary hard, 79% struggled with ChatGPT's speedy talks, 50% found it hard to understand ChatGPT's long turns. To make the chatbot's input more comprehensible, teachers might need to revise the prompt engineering or adjust ChatGPT's settings by emphasizing shorter responses or a slower rate. In addition, the teaching tasks should be revised to be more interactive, and students need to have more chances to practice negotiation for meaning.

As reported in the participants' reflection, ChatGPT adjusted its language to make the input comprehensible in different ways, with more formatting and elaborative moves than form simplification, as shown in Table 3. Most frequently, ChatGPT repeated what it said earlier, then adjusted its speech rate (i.e., speaking more slowly). Paraphrasing, providing definitions for unfamiliar words, and explaining/ giving examples were also commonly used by ChatGPT to elaborate its input. With less frequency, ChatGPT used simplification techniques such as shorter sentences, simpler grammatical structures, breaking long sentences into chunks, and using shorter words in each turn during the conversations.

These results indicate that the chatbot tended to modify its language by adjusting its delivery, such as a slower rate and repetition. This aligns with the results reported in the earlier part that the speed rate was the biggest issue that impeded participants' comprehension of the input. Furthermore, elaborative techniques (i.e., paraphrasing, defining, giving examples) were used more often than form simplification moves.

**Table 3** ChatGPT's use of modifying input techniques

| Modifying input items | N | Mean | SD | % Low (1–2) (%) | % Mid (3) (%) | % High (4–5) (%) |
|---|---|---|---|---|---|---|
| 1. Use shorter sentences | 114 | 3.07 | 0.91 | 26 | 46 | 27 |
| 2. Use simpler grammar structures | 114 | 3.09 | 0.89 | 23 | 49 | 27 |
| 3. Paraphrase | 114 | 3.32 | 0.89 | 18 | 38 | 44 |
| 4. Define difficult words | 114 | 3.30 | 0.93 | 18 | 39 | 43 |
| 5. Explain/example | 114 | 3.26 | 0.97 | 20 | 37 | 43 |
| 6. Translate into L1 | 114 | 2.05 | 1.16 | 68 | 22 | 11 |
| 7. Chunk | 114 | 2.96 | 1.04 | 32 | 39 | 28 |
| 8. Adjust turn length | 114 | 2.94 | 1.11 | 32 | 39 | 29 |
| 9. Adjust speed rate | 114 | 3.44 | 1.07 | 18 | 30 | 52 |
| 10. Repeat | 114 | 3.75 | 1.07 | 13 | 23 | 64 |

Especially, the chatbot hardly switched the input into learners' L1 in case it was prompted. This fact shows that students were engaged in conversations with a virtual partner who has rich input and is also flexible to switch between languages to help students understand when necessary. This seems to make a chatbot an even more superior conversation partner than a real native speaker, who is usually not good at the learners' mother tongue.

## 4.2 *Learners and ChatGPT's Uses of Negotiation-for-Meaning Strategies*

Figure 1 shows the perceived frequency of miscommunication between students and the chatbot during their conversations. More than half of the students reported that they sometimes experienced conversation breakdowns, and about one-fifth of them either reported seldom or often encountering conversation breakdowns, while only a very small number reported either always or never experiencing them. Given that all the students were guided on prompt engineering for the chatbot to understand how to perform communicative tasks, what could be the factors causing their differing communication failures with the AVC? Figure 2 presents the participants' responses to both chatbot-related and learner-related factors leading to communication breakdowns in chatbot–human interaction. From the chatbot side, ChatGPT's fast speed (65.8%), long turns (41.2%), and use of new words (42.1%) were the most critical factors that made the learners overwhelmed or reduced the comprehension levels of the learners during their interaction with the chatbot. Interestingly, given that their virtual conversation partner used an automatically generated voice, which could sometimes sound strange or unclear to language learners, only a very small percentage of the participants (less than 5%) reported ChatGPT's pronunciation as the cause of communication breakdowns. This seems to suggest that an AI voice chatbot could provide intelligible pronunciation to language learners. On the other hand, communication breakdowns also resulted from learners' limited knowledge of the topic (61.4%), poor pronunciation (57.9%), and inappropriate vocabulary use (42.1%), confusing structures (33.3%), and too soft voice (27.3%).

Diving into chatbot–human conversations, we explored how learners and ChatGPT dealt with communication breakdowns to keep the conversations going. Data was taken from 5 conversations between five random students with ChatGPT in an information gap task, which required the speakers to exchange information about their eating habits. Table 4 shows the total number of C-units that each student and their partner (ChatGPT) produced to complete the information gap task. The number of C-units produced was different across the five different conversations, although they were assigned the same information gap task with ChatGPT. The number of C-units produced in conversation 2 was the highest (548), followed by conversation 5 (501) and conversation 1 (440). In these three conversations, the students and their virtual partners made more than double the interactional moves than in conversations 2 and 3. Furthermore, in all cases, ChatGPT always produced far more C-units than

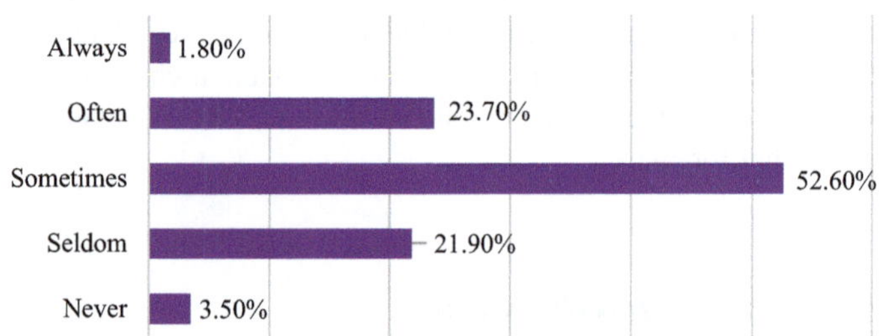


**Fig. 1** Frequency of communication breakdowns (number of participants = 114)

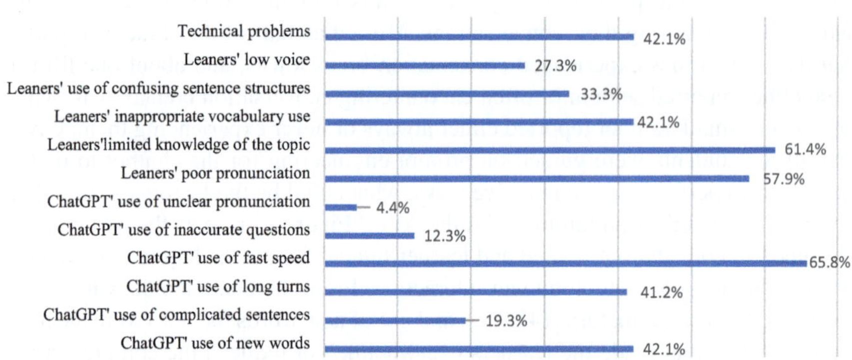


**Fig. 2** Causes of communication breakdowns in human–chatbot interaction

the learners, which also indicates the longer turns ChatGPT spoke. In other words, ChatGPT appeared to dominate the conversations and give more information than did the students. The results were similar to Foster's (1998) findings in classroom interaction, which highlighted the domination of one speaker over another in a dyad.

To explore the negotiation of meaning in the utterances used by each speaker, the C-units were categorized into four groups of meaning negotiation moves (e.g., confirmation checks, comprehension checks, clarification requests, and repetition/ speed adjustment requests). The numbers of moves were counted by each speaker in each conversation of the same information gap task, as shown in Table 5. Students produced far more negotiation of meaning moves than ChatGPT, with 121 moves compared to 35. This suggests that students struggled to comprehend ChatGPT's input and managed to keep communication flowing by using techniques for meaning negotiation. In the students' C-units, the distribution of move types also reveals that they mainly requested clarification by asking for definitions, paraphrases, or examples, as well as repetition or speed requests. In contrast, ChatGPT mostly used confirmation checks (57% of its moves), occasionally comprehension checks, and rarely clarification or repetition requests.

Comparing with the total number of C-units produced by each speaker in the whole conversation, Table 6 shows that the students used far more meaning negotiation strategies than ChatGPT). While the number of meaning negotiation moves from students was quite high, the number used by ChatGPT was very low. For example, speaker 1 used 27 C-units out of 167 (16%) for meaning negotiation. That proportion remained similar for speakers 3 (40%) and speaker 4 (32%). Speakers 2 and 3 negotiated for meaning less frequently, with 5% and 10% respectively. Some sample utterances of meaning negotiation included "*Is that right?*" (S2), "*Can you repeat that*?" (S2), or "*Again*!" (S4), "*Cool, right?*" (S2), "*Slow down*!" (S1), "*How do you spell PASTA*?" (S2). In contrast, ChatGPT's incidences of meaning negotiation were very low, ranging from 1 to 4%. These results indicate that ChatGPT could understand what speakers said easily without the need for further clarification or repetition requests or comprehension checks. It sometimes checked the confirmation to ensure that the information it received was correct. Some sample questions that ChatGPT used to negotiate meaning as "*So, you mean you tend to have some more snacks?*" (conversation 2), "*So just to make sure I got you right … ?*" (conversation 3).

The unbalanced number of meaning negotiation moves between two speakers in the same dialogue (human and ChatGPT), and the unpopular distribution of these strategies in chatbot–human conversation, were in line with those in human–human interaction in the classroom, as stated in Foster's (1998) findings. Negotiation in interactional communication was rarely shared equally among participants, and it was often driven by one or two active individuals.

Overall, the results indicate that students and ChatGPT negotiate for meaning in complementary but unequal ways: students actively initiate clarification and repair, while ChatGPT mainly confirms interpretations. As Foster (1998) argued, this pattern shows that negotiation is fragile in classroom-like contexts and unlikely to lead to deeper language modification unless tasks are tightly designed or students are explicitly trained to pursue extended negotiation routines.

**Table 4** The total number of C-units provided by students and ChatGPT in conversations

| Speakers | Conversation 1 | | Conversation 2 | | Conversation 3 | | Conversation 4 | | Conversation 5 | |
|---|---|---|---|---|---|---|---|---|---|---|
| | S1 | ChatGPT | S2 | ChatGPT | S3 | ChatGPT | S4 | ChatGPT | S5 | ChatGPT |
| Numbers of C-units by speakers | 167 | 273 | 150 | 398 | 65 | 139 | 50 | 167 | 177 | 324 |
| Total number of C-units | 440 | | 548 | | 204 | | 217 | | 501 | |

**Table 5** Distributions of meaning negotiation moves

| | Conversation 1 | | Conversation 2 | | Conversation 3 | | Conversation 4 | | Conversation 5 | |
|---|---|---|---|---|---|---|---|---|---|---|
| | S1 | ChatGPT | S2 | ChatGPT | S3 | ChatGPT | S4 | ChatGPT | S5 | ChatGPT |
| 1. Confirmation checks | 4 | 8 | 6 | 1 | 6 | 1 | 0 | 4 | 3 | 6 |
| 2. Comprehension checks | 1 | 0 | 4 | 3 | 3 | 0 | 0 | 1 | 1 | 2 |
| 3. Clarification requests (incl. asking for example/reason/ definition/paraphrase/ example/translation requests) | 4 | 2 | 19 | 0 | 13 | 0 | 15 | 0 | 14 | 2 |
| 4. Repetition/speed requests | 18 | 0 | 5 | 0 | 4 | 5 | 1 | 0 | 0 | 0 |
| Total | 27 | 10 | 34 | 4 | 26 | 6 | 16 | 5 | 18 | 10 |

**Table 6** Distributions of students' modified output moves

| | S1 | S2 | S3 | S4 | S5 |
|---|---|---|---|---|---|
| Semantic | 2 | 9 | 1 | 2 | 4 |
| Morphological | 1 | 4 | 1 | 1 | 2 |
| Phonological | 1 | 3 | 0 | 1 | 1 |
| Syntactic | 2 | 6 | 1 | 1 | 2 |
| Total | 6 | 22 | 3 | 5 | 9 |
| Percentage per total C-units (%) | 3.6 | 14.6 | 4.6 | 10 | 5.1 |

## *4.3 Students' Modification of Their Language Output During Their Conversation with ChatGPT*

To examine how students modified their output while interacting orally with ChatGPT, we categorized the C-units produced by the 5 students into four categories as operationalized by Pica et al. (1989). First, semantic modifications include the use of synonyms, paraphrases, and examples. The second category is morphological modifications, including adding, deleting, or replacing function words or word endings. The next category is phonological modification, which refers to the changes in pronunciation for better clarity. Finally, the syntactic modifications cover the restructuring and elaboration of clauses.

Table 6 shows a rare distribution of modifying output among the five students in the information exchange gap activity. The students mostly modified their language use, changing the words, using synonyms, paraphrases, and changing word use (semantic modification). For instance, student 5 self-corrected her word usage after noticing the mistake: "I try to keep myself to have a healthier habit." Following that, syntactic modification occurred from time to time by correcting the use of sentence structures after getting explicit correction from ChatGPT. For example, student 5 corrected the use of present simple tense verbs after listening to how ChatGPT corrected the mistakes.

*Student 5: But I am not very like eating.*

*ChatGPT: Ah you should use an auxiliary verb, not to be, "I don't really like eating."*

*Student 5: Yes, sorry. "I* ***don't really****like eating."*

Students did not tend to change the ending of the words (morphological modification) very often. Atypical example was produced by speaker 1 as follow:

*Student 1: I eat three meal a day.*

*ChatGPT: "You should be aware of singular and plural nouns. Like three meals instead of three meal."*

*Student 1: Oh, yes. Three* ***meals****a day.*

Finally, phonological changes were the least common type of modification that the students made. Even though learners made a lot of pronunciation mistakes, they did not seem to notice or correct them. Alternatively, ChatGPT did not give feedback on this aspect, but it often tried to interpret the learners' language successfully. Therefore, students did not have a chance to modify their pronunciation. A sample piece of dialogue between student 1 and ChatGPT demonstrated how she modified the pronunciation after ChatGPT's confirmation check.

*Student 1: "Do you refer fly or heavy flint?"*

*ChatGPT: Do you mean light or heavy lunch?*

*Student 1: Yes, that's right. Do you prefer* ***light or heavy****lunch?*

When comparing the distribution of output modifications among the five students, it is evident that not all students had the same frequency of modifying their outputs. For example, S2 made 22 moves of the total C-units he produced, and S3 made 9 changes, while S1 modified 6 times, and student 4 made 5 changes. Student 3 made the fewest output modification moves (3 times).

When we compare these results with Foster's (1998) study, we can see the similarities between output modification in human–human interaction and human-chatbot interaction. Like Foster's learners, the most common type of output modification here was semantic, and the work was done by only a few active students. In addition, syntactic, morphological, and phonological changes were very rare. This suggests that interacting with ChatGPT may not have encouraged students to adjust their language use, just as when they interacted with peers. Still, the uneven pattern across learners confirms Foster's point that modified output is not equally shared and that many students remain passive unless tasks or prompts explicitly push them to make changes.

## 5 Conclusion and Implications

In summary, this chapter has demonstrated the potential of using such AI voice-based chatbots as ChatGPT in the setting of Vietnamese English learning classes. The analysis of how ChatGPT and learners adjusted their linguistic use to negotiate meaning reinforced the central claim that AVCs and students can sustain interaction with different strategies to make their language comprehensible to one another. Our findings extend prior work on chatbot–human interaction by shedding light on the way both learners and ChatGPT have adjusted their language to get the message across, thereby maintaining the conversation flow. With these findings, the study is among the first of its kind to explore an uncharted territory of how language learners can orally interact with AI voice chatbots. It enriches the body of literature on how human-chatbot interaction takes place in a class setting, elucidating the process of meaning construction as well as communication breakdown repair in AI-mediated contexts. The results of the study help to shed more light on the limited body of

Vietnamese EFL research on using AVCs for speaking skill development, connecting to and complementing the findings from earlier studies by Nguyen (2025) and Duong and Suppasetseree (2024).

The implications for using AVCs for speaking practice are significant both theoretically and practically. First, this study is likely to be among the early attempts to examine the chatbot–human interaction through the lens of the Interaction Hypothesis, which traditionally focuses on human–human interaction. It might add to the theory that both chatbots and learners could interact with each other orally as conversation partners to complete speaking tasks. For example, ChatGPT can adjust its language to make the input comprehensible. Students could negotiate for meaning when miscommunication occurred by techniques like making clarification requests, confirmation checks, speed adjustment, and repetition requests, and keep the conversations going. During the interaction, students could modify their outputs, showing that they could understand the input and tried to use it in their turns, although it only occurred at a moderate frequency. These insights might set the ground for future studies to refine the Interactional Hypothesis in the chatbot–human interaction context. For practical implications, the results suggest that it is important to provide the learners with training such as prompt engineering, negotiation-of-meaning strategies, and chatbot operation before their experience. Moreover, to foster more vivid interactional moves between the human and chatbot partner, teachers need to design the speaking tasks that require more negotiation of meaning to occur naturally. For each speaking task, it is also vital to provide the chatbot with a clear role, specific prompts, and feedback types to maximize the chatbot–human interactional dynamics.

While this study contributes to a better understanding of learner-chatbot interaction in classroom settings, it is not without its limitations. First, our study only explored chatbot–human interaction with a focus on participants' perception and recording analysis. Future research might conduct experiments to examine the impacts of this chatbot-integrated speaking task on learners' speaking skills. For example, future researchers might compare two groups of students (i.e., human–human and chatbot–human interactions) under controlled communicative tasks such as role play, discussion, or even information gaps. To provide interesting insights into how chatbot–human interaction affects students' speaking outcomes. Second, the number of students is still small ($N = 114$ for survey, $N = 5$ for screen recording), which may not fully reflect the actual interaction of chatbot–human interaction in students of different levels of language proficiency. Thirdly, our study only focuses on exploring the chatbot–human interaction in the light of three core constructs of Long's (1996) IH, which may neglect the corrective feedback and learners' attention to the linguistic features during their conversation with ChatGPT. Future studies should delve into these dimensions of the Interaction hypothesis to have a better understanding of the dynamics of how chatbots give corrective feedback and what types of linguistic items students might notice to turn the input into intake.

**Author Contributions Statement** Ho Thi Hien (first author): led the conceptualization and design. She conducted the literature review, data collection, analysis and interpretation, drafted the manuscript and integrated co-authors' feedback.

Nguyen Van Huy (correspondent author): supervised the study, critically revised the manuscript, contributed to design, and methodology. He also reviewed analyses, discussed findings and conclusion.

Nguyen Thi Bich Hanh: Contributed to methodology, inter-coding data, data analysis and interpretation, proofreading and editing the manuscript.

All authors approved the final version of the manuscript and agree to be accountable for all aspects of the work.

**Acknowledgement** This work was partly supported by research funding from Hue University and the Industrial University of Ho Chi Minh City.

**Disclosure of AI Writing Assistance** We employed ChatGPT's assistance regarding language editing such as sentence clarity, grammar and coherence improvement for the first draft. NotebookLM was also used for transcribing the recording and initially coding the segments of text into C-units. All the AI assistance was verified and revised by the researchers.

**Conflict of Interest** There are no conflicts of interest related to this work among the authors.

## References

Adamopoulou, E., & Moussiades, L. (2020). An overview of chatbot technology. In I. Maglogiannis, L. Iliadis, & E. Pimenidis (Eds.), *Artificial intelligence applications and innovations: AIAI 2020, IFIP advances in information and communication technology* (Vol. 584, pp. 373–383). Springer. https://doi.org/10.1007/978-3-030-49186-4_31

Al-Tamimi, N., Abudllah, N., & Bin-Hady, W. R. A. (2020). Teaching speaking skill to EFL college students through task-based approach: Problems and improvement. *British Journal of English Linguistics, 8*(2), 113–130.

Azizimajd, H. (2023). Investigating the impacts of voice-based student-chatbot interactions in the classroom on EFL learners' oral fluency and foreign language speaking anxiety. *Technology Assisted Language Education, 1*(2), 61–83. https://doi.org/10.22126/tale.2023.2732

Bashori, M., van Hout, R., Strik, H., & Cucchiarini, C. (2024). I can speak: Improving English pronunciation through automatic speech recognition-based language learning systems. *Innovation in Language Learning and Teaching, 18*(5), 443–461. https://doi.org/10.1080/17501229.2024.2315101

Belda-Medina, J., & Calvo-Ferrer, J. R. (2022). Using chatbots as AI conversational partners in language learning. *Applied Sciences, 12*(17), Article 17. https://doi.org/10.3390/app12178427

Brown, H. D. (2001). *Teaching by principles: An interactive approach to language pedagogy.* Addison Wesley Longman.

Çakmak, F. (2022). Chatbot-human interaction and its effects on EFL students' L2 speaking performance and anxiety. *Novitas-ROYAL (Research on Youth and Language), 16*(2), 113–136.

Channell, M. M., Loveall, S. J., Conners, F. A., Harvey, D. J., & Abbeduto, L. (2018). Narrative language sampling in typical development: Implications for clinical trials. *American Journal of Speech-Language Pathology, 27*(1), 123–135. https://doi.org/10.1044/2017_AJSLP-17-0046

Creswell, J. W., & Clark, V. L. P. (2018). *Designing and conducting mixed methods research.* Sage Publications.

Ding, D., & Yusof, M. B. (2025). Investigating the role of AI-powered conversation bots in enhancing L2 speaking skills and reducing speaking anxiety: A mixed methods study. *Humanities and Social Sciences Communications, 12*(1), 1–16. https://doi.org/10.1057/s41599-025-05550-z

Doughty, C., & Pica, T. (1986). "Information gap" tasks: Do they facilitate second language acquisition? *TESOL Quarterly, 20*(2), 305–325.

Du, J., & Daniel, B. K. (2024). A Systematic review of AI-powered chatbots in EFL speaking practice: Transforming language education. *Computers and Education: Artificial Intelligence*. https://doi.org/10.1016/j.caeai.2024.100230

Duong, T., & Suppasetseree, S. (2024). The effects of an artificial intelligence voice chatbot on improving Vietnamese undergraduate students' English speaking skills. *International Journal of Learning, Teaching and Educational Research, 23*(3), 293–321. https://doi.org/10.26803/ijlter.23.3.15

Fathi, J., Rahimi, M., & Derakhshan, A. (2024). Improving EFL learners' speaking skills and willingness to communicate via artificial intelligence-mediated interactions. *System, 121*. https://doi.org/10.1016/j.system.2024.103254

Foster, P. (1998). A classroom perspective on the negotiation of meaning. *Applied Linguistics, 19*(1), 1–23.

Foster, P., Tonkyn, A., & Wigglesworth, G. (2000). Measuring spoken language: A unit for all reasons. *Applied Linguistics, 21*(3), 354–375.

Gan, Z. (2012). Understanding L2 speaking problems: Implications for ESL curriculum development in a teacher training institution in Hong Kong. *Australian Journal of Teacher Education (Online), 37*(1), 43–59. https://doi.org/10.14221/ajte.2012v37n1.4

Hayden, M., & Le Tran, T. (2024). *Education in Vietnam: Making haste slowly*. Taylor & Francis.

Hunt, K. W. (1966). Recent measures in syntactic development. *Elementary English, 43*(7), 732–739.

Hwang, G. J., Rahimi, M., & Fathi, J. (2024). Enhancing EFL learners' speaking skills, foreign language enjoyment, and language-specific grit utilising the affordances of a MALL app: A microgenetic perspective. *Computers & Education, 214*, 105015. https://doi.org/10.1016/j.compedu.2024.105015

Jeon, J. (2024). Exploring AI chatbot affordances in the EFL classroom: Young learners' experiences and perspectives. *Computer Assisted Language Learning, 37*(1–2), 1–26. https://doi.org/10.1080/09588221.2021.2021241

Kalimullah, M., & Tabassum, F. (2025). Technology-supported English language learning: Effectiveness and limitations. *International Journal of Social Impact, 10*(1), 26. https://doi.org/10.25215/2455/1001026

Koç, F. Ş, & Savaş, P. (2025). The use of artificially intelligent chatbots in English language learning: A systematic meta-synthesis study of articles published between 2010 and 2024. *ReCALL, 37*(1), 4–21. https://doi.org/10.1017/S0958344024000168

Krashen, S. (1982). *Principles and practice in second language acquisition*. Pergamon Press.

Le, V. T., & Thi, D. H. (2022). Factors affecting English language teaching in Vietnamese high schools from teachers' perspective. *International Journal of English Literature and Social Sciences, 7*(1), 317–328. https://doi.org/10.22161/ijels.71.44

Lim, Y. (2024). Learning effects of complementary learning methods in speaking skills using AI-supported devices. *The Journal of Foreign Studies, 68*(6), 91–112. https://doi.org/10.15755/jfs.2024.68.91

Liu, Y., binti Ab Rahman, F., & binti Mohamad Zain, F. (2025). A systematic literature review of research on automatic speech recognition in EFL pronunciation. *Cogent Education, 12*(1), 2466288. https://doi.org/10.1080/2331186X.2025.2466288

Loban, W. (1976). *Language development: Kindergarten through grade twelve* (NCTE Committee on Research Report No. 18). National Council of Teachers of English.

Long, M. H. (1996). The role of the linguistic environment in second language acquisition. In W. C. Ritchie & T. K. Bhatia (Eds.), *Handbook of second language acquisition* (pp. 413–468). Academic Press.

Mackey, A., & Gass, S. M. (2015). *Second language research: Methodology and design*. Routledge.

McAleavy, T., Ha, T. T., & Fitzpatrick, R. (2018). *Promising practice: Government schools in Vietnam*. Eric.

Muniandy, J., & Selvanathan, M. (2025). ChatGPT, a partnering tool to improve ESL learners' speaking skills: Case study in a public university, Malaysia. *Teaching Public Administration, 43*(1), 4–20.

Nguyen, T. H. (2025). The impact of AI chatbot integration on Vietnamese EFL learners' speaking performance, fluency, and confidence. *Language, Technology, Social Media, 4*(1), 70–88. https://doi.org/10.70211/ltsm.3026-7196.314

Nguyen, T. P. L., & Burke, R. (2025). Foreign language teaching in Vietnam: Opportunities and challenges. In L. T. Tran, & S. Marginson (Eds.), *Education in Vietnam* (pp. 145–158). Bloomsbury Publishing. https://doi.org/10.4324/9781003298700-11

Nunan, D. (2003). *Practical English language teaching.* McGraw-Hill Education.

OpenAI. (2025a). *Introduction GPT-5.* OpenAI. https://openai.com/vi-VN/index/introducing-gpt-5/

OpenAI. (2025b). *Voice mode FAQ.* https://help.openai.com/en/articles/8400625-voice-chat-faq

Pavlik, J. V. (2023). Collaborating with ChatGPT: Considering the implications of generative artificial intelligence for journalism and media education. *Journalism & Mass Communication Educator, 78*(1), 84–93. https://doi.org/10.1177/10776958221149577

Pérez, J. Q., Daradoumis, T., & Puig, J. M. M. (2020). Rediscovering the use of chatbots in education: A systematic literature review. *Computer Applications in Engineering Education, 28*(6), 1549–1565. https://doi.org/10.1002/cae.22326

Pham, K. T., Ha, X. V., Tran, N. H., & Nguyen, Y. T. X. (2023). Curriculum reform in Vietnam: Primary teachers' views, experiences, and challenges. *Education 3–13, 51*(3), 440–451. https://doi.org/10.1080/03004279.2022.2162829

Pham, V. T. T., & Pham, A. T. (2025). English major students' satisfaction with ELSA speak in English pronunciation courses. *PLoS ONE, 20*(1), e0317378. https://doi.org/10.1371/journal.pone.0317378

Pica, T., Holliday, L., Lewis, N., & Morgenthaler, L. (1989). Comprehensible output as an outcome of linguistic demands on the learner. *Studies in Second Language Acquisition, 11*(1), 63–90. https://doi.org/10.1017/S027226310000783X

Sun, W. (2023). The impact of automatic speech recognition technology on second language pronunciation and speaking skills of EFL learners: A mixed methods investigation. *Frontiers in Psychology, 14*, 1210187. https://doi.org/10.3389/fpsyg.2023.1210187

Swain, M. (1985). Communicative competence: Some roles of comprehensible input and comprehensible output in its development. *Input in Second Language Acquisition, 15*, 165–179.

Tanka, J., & Most, P. (2020). *New interactions 2.* McGraw Hill Press.

Ur, P. (1996). *A course in language teaching* (Vol. 1). Cambridge University Press.

Yang, H., Kim, H., Lee, J. H., & Shin, D. (2022). Implementation of an AI chatbot as an English conversation partner in EFL speaking classes. *ReCALL, 34*(3), 327–343. https://doi.org/10.1017/S0958344022000039

Yufrizal, H. (2001). Negotiation of meaning and language acquisition by Indonesia EFL learners. *TEFLIN, 12*(1), 60–87. https://doi.org/10.15639/teflinjournal.v12i1/60-87

Zou, B., Wang, C., Yan, Y., Du, X., & Ji, Y. (2025). Exploring English as a foreign language learners' adoption and utilisation of ChatGPT for speaking practice through an extended technology acceptance model. *International Journal of Applied Linguistics, 35*(2), 689–704. https://doi.org/10.1111/ijal.12658

**Ho Thi Hien** is a lecturer of English at the Industrial University of Ho Chi Minh City, Vietnam. She is a Ph.D. student in TESOL at the University of Foreign Languages and International Studies, Hue University. Her research interests include technology-enhanced language learning (TELL), learners' differences, and intercultural communicative competence (ICC).

**Nguyen Van Huy** is a lecturer in Applied Linguistics and TESOL at the University of Foreign Languages and International Studies, Hue University, Vietnam. He earned his Ph.D. in Language Education from the University of Queensland, Brisbane, Australia. His main research interests lie in language planning and policy (LPP) with a special focus on language-in-education policy, English as a Medium of Instruction (EMI), and educational leadership in developing countries.

**Nguyen Thi Bich Hanh** is Head of English Department at Ha Tinh University, Vietnam. Her current research interests include text-level reading comprehension, application of technology in L2 language teaching, and ESL teaching material development and evaluation. She received a Fulbright-funded master's degree in TESOL from the US and an Erasmus Mundus doctoral degree in Brain and Language Science from the UK and Germany.

# EFL Learners' Engagement in ChatGPT–Generated Speaking Tasks from a Text-Driven Approach: A Case in Vietnam

**Huynh Truong Sang and Nguyen Van Loi**

**Abstract** Individual differences–specifically task engagement (TE)–in task-based language teaching have recently stimulated scholarly inquiry among second language researchers and educators. Nevertheless, most studies focus on learners' engagement in human-designed tasks; few studies have employed ChatGPT as a tool to develop task-based materials to facilitate language learning. This study aimed to narrow the gap by exploring how Vietnamese English majors engaged with tasks generated by ChatGPT in accordance with a text-driven task-based approach (Tomlinson, Brian (ed), Developing materials for language teaching, Bloomsbury Publishing, 2023). Two 50-min speaking task-based lessons generated by ChatGPT based on preselected texts were delivered to thirty English sophomores at a public university. An explanatory sequential design was used to gather data, including a questionnaire on task engagement after each lesson, and a follow-up semi-structured interview to delve into dimensions of engagement. Findings showed that participants were highly engaged in the speaking tasks, consistently in four dimensions across the two lessons. Behaviourally, the learners tried to complete and stay focused on their tasks despite challenges; cognitively, they attempted to connect relevant information for task completion, while emotional engagement was reflected in their interest in trendy topics while participating in role-play and opinion exchanges. Social engagement involved seeking help with new words and ideas for speaking. The study implies that teachers can use AI for supporting task design and implementation to promote learner engagement.

**Keywords** ChatGPT · EFL · Task engagement · Text-driven tasks · Vietnam

H. T. Sang · N. Van Loi (✉)
School of Foreign Languages, Can Tho University, Can Tho, Vietnam
e-mail: loinguyen@ctu.edu.vn

H. T. Sang
Faculty of Foreign Languages, An Giang University, Long Xuyen, Vietnam

Vietnam National University Ho Chi Minh City, Ho Chi Minh City, Vietnam

 
H. P. Bui and L. P. H. Huong (eds.), *Artificial Intelligence in Education and Academic Research*, Studies in Computational Intelligence 1269,
https://doi.org/10.1007/978-3-032-18392-7_7

# 1 Introduction

TE has recently attracted a growing body of research as a prerequisite for facilitating task performance and language development (Aoyama et al., 2024; Philp & Duchesne, 2016). Studies on TE have largely been situated within a cognitive approach, highlighting the impacts of various factors from task types, topics, choices, interaction, and implementation models on learners' engagement. These studies have primarily drawn on the cognitive dimension of task as input for task design while affective factors that are purposefully designed to drive TE have received little attention.

Tomlinson (2018, 2023) proposed a text-driven approach which deploys emotionally and cognitively engaging texts as a meaningful resource for task design and implementation to affect learners' engagement and facilitate language production. Taking text-related variables into account in task design in this way echoes Ellis' (2024) proposal about the interaction between task factors (i.e., design and implementation) and learners' engagement during task performance in affecting their language production and task outcomes. Richards (2022) posited three influential factors on learner emotional engagement such as materials (i.e., texts), teachers, and peers. Aubrey et al. (2022) further found that the overall lesson level, learners, and task factors served as contributory factors to learners' TE.

ChatGPT, as an artificial intelligence (AI) Chatbot, has recently emerged and been used in various fields, including language education. AI applications have been found to exert positive influences over learners' language skills and knowledge (Fathi et al., 2024; Karataş et al., 2024; Pan et al., 2024), and individual difference factors such as motivation, autonomy, enjoyment, emotions, and engagement (Xin & Derakhshan, 2025; Yuan et al., 2025; Zhang & Miao, 2025; Zhou & Hou, 2024). Among these, ChatGPT's effects on learners' engagement in language learning, particularly regarding task performance, have been under-researched (Lai & Sundqvist, 2025; Lo et al., 2024). While opportunities and affordances of AI have been proposed for optimising and enhancing EFL teaching materials (Singha et al., 2024), research on exploiting AI for generating instructional materials to impact learners' engagement has been scarce. In this context, following Tomlinson's (2018) principled guidelines for materials development, this study thus sought to explore ChatGPT as a support tool in designing and implementing speaking task-based lessons to affect learners' engagement. The following question is explored:

*What is Vietnamese EFL learners' engagement in text-driven speaking tasks generated by ChatGPT?*

## 2 Literature Review

### 2.1 A Text-Driven Approach to Task-Based Instruction (TDTB)

Tomlinson (2013) outlines a text-driven approach (TDA) that emphasises the selection of cognitively and affectively engaging texts as a central component in developing meaningful activities for language learning. According to the author, the TDA is a comprehensive pedagogical framework for developing materials for language teaching which particularly uses engaging texts to prompt learners' cognitive and affective engagement, which in turn fosters their language development. To achieve this, Tomlinson presents a pedagogical framework that involves a series of meaningful activities leading learners step by step to connect with, experience, react to and deploy selected texts for language learning. These activities are grounded in second language acquisition principles, including personal connections, visualisation, inner speech, interaction, cognitive and affective engagement, and noticing.

Tomlinson (2018, 2023) then developed the TDTB framework which incorporates the text-driven principles into task-based language teaching from which cognitive and affective text engagement is considered for task design and execution. Arguing that learners may use text inputs for completing tasks, Tomlinson characterises task performance as being influenced by potentially engaging texts that are preselected in reference to specific criteria (Tomlinson, 2023). According to the author, the TDTB approach offers several benefits such as upholding learners' emotional engagement (negative and positive emotions), and cognitive engagement (think to solve the problem in the text, connect to their personal experiences, understand values of the text), generating meaningful learning through the integration of authentic content that might be excluded from the course syllabus, creating more like real-world tasks which stimulate learners' interest, and encouraging exposure to rich and meaningful input. The TDTB procedure is described as follows (Tomlinson, 2018, p. 6; Tomlinson, 2023).

Step 1: A readiness activity is intended to draw learners' attention to the subject, topic, or setting of the text and task. Learners' prior knowledge is activated through one of the following activities: visualisation while viewing photos or a video, personal connection to their own life experiences, prediction, or a mini-role play.

Step 2: This so-called text experience stage stimulates learners to read the text for personal meaning, and not the linguistic aspects used to express the meaning. They can be asked to visualise, check their predictions (i.e., check what they predicted in step 1 in connection to the information collected from the text), or internally interrogate the writer or speaker.

Step 3: This stage aims at deepening learners' personal interpretation and interaction with the text, by linking their initial thinking with their thinking back to the text, expressing emotions about the material, addressing a controversial statement from the text, or summarising it to others. This activity is implemented after learners have been familiar with the text.

Step 4: A text-driven task is then provided to engage learners in expressing their personal perspectives, for example, writing a letter to the text author (i.e., expressing their opinions about their preferred information from the text), expanding the text with their own information or expression which is coherently connected to the available text content, presenting an initiative, or providing a viable solution to the problem explicitly or implicitly posed in the text. Like the task phase in TBLT, this step is crucial to the TDTB approach, but text-based interaction is encouraged during task performance. Learners can together study and respond to the task demands to complete the task. Ideas generated from each task performer after the reading might contribute to their successful task performance in a low cognitively loaded manner. This makes a text-driven task distinct from a cognitively directed task which requires learners to refer to their own repertoires without reference to the meaningful text input.

Step 5: In this stage, learners are given time to observe each other's text-driven task performance and give feedback to better its quality. This peer monitoring activity allows learners to collaboratively work in pairs or groups to give advice on their task performance.

Step 6: Learners compare their performance with a proficient L2 learner's performance, with a preferred focus on a particularly salient linguistic feature (e.g., lexical, pragmatic) of the text.

Step 7: Learners revise the task or perform a different but similar task.

Step 8: Learners are further prompted to search for texts that contain similar language samples studied in Step 6.

It is noteworthy, according to Tomlinson, that (1) a unit does not need to involve every step; the sequence of steps 5–8 can be changed; and (2) teachers should be flexible in their response to the activity during step 4. Apart from the text-related contribution, a text-driven task should follow four features as proposed by Ellis (2009).

## 2.2 *Task Engagement*

Tasks as a basic unit of teaching can be real-world and pedagogic. Pedagogic or classroom tasks are intended to promote learner language use in a freer manner, allowing them to be involved in meaning negotiation to arrive at task outcomes. These tasks should contain four features described by Ellis (2009): (1) a meaning-oriented focus, (2) an information gap for learners to fill in (share information or opinions), (3) a clearly-defined outcome with a focus on using target language to achieve a communicative goal, not just practice language itself, and (4) opportunities for learners to use their own resources to complete the task. All of these make a task different from a grammatical exercise which is restricted to predetermined language items.

Learners' engagement generally denotes "how actively involved a student is in a learning task and the extent to which that physical and mental activity is goal-directed and purpose-driven" (Hiver et al., 2024, p. 203). TE, derived from such a broader concept, is a prediction variable for successful task execution or performance (Dao, 2021; Hiver & Wu, 2023). Ellis (2024) posits TE can be influenced by task factors (design and implementation), teacher factors (teacher choices of task types and task sequence), learner factors (individual differences in knowledge both non-linguistic and linguistic), and contextual factors (whether the tasks are performed in face-to-face or tech-mediated environments).

Fredricks et al. (2004) described engagement in triangular dimensions: behavioural, cognitive, and affective. Expanding this construct, Philp and Duchesne (2016, p. 51) later defined it as "a state of heightened attention and involvement," behaviourally, emotionally, cognitively, and socially. Each dimension encompasses distinct indicators that interact and influence one another. Behavioural engagement is primarily concerned with the number of turn-taking in collaboration, duration of spontaneous runs, and time-on-task investment (Lambert, 2017; Nakamura et al., 2021). Cognitive involvement is described as psychological endeavour, readiness, and attentional focus in the learning process (Svalberg, 2018). Emotional or affective engagement pertains to learners' emotional experiences during task performance (negative and positive). Social engagement, particularly during tasks, indicates learners' well-preparedness to interact, to actively get involved in interaction, to provide interlocutors with support during interaction (Dao, 2021).

This study aimed to explore EFL learners' engagement in speaking tasks driven by texts. Thus, the conceptual framework in Fig. 1 was established. Accordingly, TE was examined in terms of four dimensions, and the TDA was adapted into three phases: pre-task, task, and post-task phase.

## 2.3 *Artificial Intelligence and ChatGPT in Language Learning*

Using technology to support materials design, this study explored the deployment of ChatGPT for designing task-based lessons. Activating human cognitive abilities through programmed algorithms to perform various tasks (Son et al., 2025), AI potentially serves as an assistant directed under human prompts or commands. ChatGPT, developed based on OpenAI as part of GenAI, was built upon large language models that are trained to provide as many human-like responses as possible in 2022. The tool has garnered scant attention due to its strong affordances for L2 education, especially effective in fostering EFL learners' language production (Muniandy & Selvanathan, 2025; Songsiengchai, 2025). According to Moorhouse and Wong (2025), ChatGPT is also useful for assisting language teachers in lesson planning and materials preparation. Teachers can engage in conversation with ChatGPT, together with their own knowledge of language teaching, to develop lesson plans, which reduce their

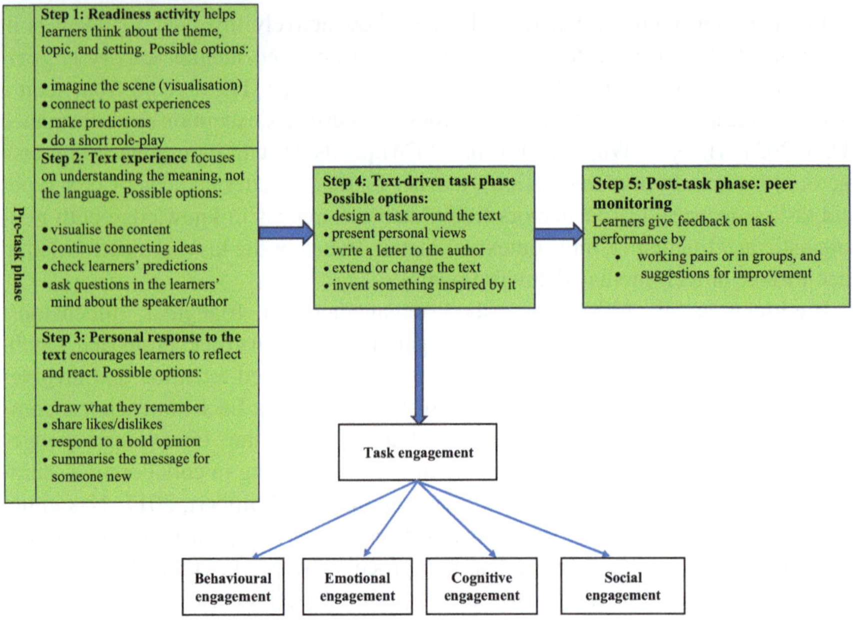


**Fig. 1** Adapted TDTB framework and dimensions of TE

workload in lesson preparation. Together with criteria for materials development (Tomlinson, 2013), teachers can potentially generate text genres (films, new articles, audio scripts, films) that have a high potential for engaging learners in language production (Moorhouse & Wong, 2025). However, in this study, ChatGPT was used to support the design of tasks with reference to prompts and input texts provided by the researchers.

## *2.4 Impacts of AI and ChatGPT on Learners' Engagement*

A body of scholarly works has discussed the role of AI in promoting learners' engagement and language learning. Using a quasi-experimental design, Ma and Chen (2024) found 350 Chinese EFL first-year students' three-dimensional engagement (i.e., behavioural, cognitive, and emotional) was improved and their academic procrastination levels were minimised when AI-supported tools such as ChatGPT, POE, and Duolingo were applied in learning. In the same context, Zhou and Hou (2024) drew on a qualitative approach to investigate engagement of 26 EFL teachers and 38 college language learners from different universities in China. Results indicated that (1) learners behaviourally engaged with AI more often, utilising AI-supported tools in language learning, (2) teachers and learners cognitively engaged with using AI in supporting their learning and teaching strategies, indicating concerns over their over-reliance (3) learners generally expressed positive emotions regardless of teachers' fear about unemployment and lower student–teacher interactions.

Teng (2024) also yielded positive results regarding ChatGPT as a feedback-generated assistant which enhanced EFL learners' motivation, engagement, social interaction, and self-efficacy during a long-term academic writing course. Using stimulated recall interviews and reflective journals, Liu and Yu (2022) found that EFL learners invested more time and made cognitive efforts in addressing automated comments generated by Write & Improve with Cambridge (WWC) system in a writing course. Similarly, Meşe et al. (2025) employed journals and semi-structured interviews to investigate learners' engagement in an online L2 writing course under ChatGPT support. The findings revealed that learners made great efforts in searching for information, feedback, error correction as indicators of behavioural engagement. Cognitively, they endeavoured to think critically and notice their knowledge gaps. They were emotionally engaged with ChatGPT, expressing fun and curiosity. Less disengagement was described at low levels of overreliance and disruption. Using a quasi-experimental design, Tran and Nguyen (2024) testified to the effects of ChatGPT as an assistant on learners' engagement. Findings illustrated that the ChatGPT-assisted group was more engaged and motivated than their counterparts. Similarly, Nguyen (2025) reported on the high engagement level in AI-supported speaking practice among Vietnamese undergraduate students.

Overall, previous studies mainly examined learners' engagement with using AI for language learning, while studies on using AI as an assistant in materials development are neglected. They also present a gap in exploring four dimensions of TE (Meşe et al., 2025). The present study thus investigates learners' engagement in ChatGPT-generated text-driven tasks which are designed and executed in the TDTB approach (Tomlinson, 2018, 2023).

## 3 Methods

### *3.1 Design*

The present research relied on an explanatory sequential mixed methods design, combining quantitative and qualitative measures to elucidate the research problem (Creswell & Guetterman, 2024). To explore cognitive, affective, behavioural, and social dimensions of learners' TE, two instruments were employed in two consecutive stages: (1) the collection and analysis of the TE questionnaire adopted from Zare and Derakhshan (2024), and then (2) a semi-structured interview to elaborate on the questionnaire results.

## 3.2 Materials

Two 50-min lessons were designed and implemented in accordance with a five-step procedure in the TDTB approach (Tomlinson, 2018, 2023). Two main tasks were designed, corresponding to two selected texts: Task 1 is problem-solving role-play based on a news story about a Vietnamese yelling vendor and task 2 is an opinion exchange based on a presentation video about the human life cycle. ChatGPT was used to support the design, following the same structured template as follows:

**Command 1**: I (as researcher) provide you (as ChatGPT) with various sources of input below. Please wait until I have given you all of them before responding.

Input 1: The content of a cognitively and affectively engaging text selected according to Tomlinson's (2023) guided questions and judged by two lecturers.

Input 2: Five steps to design a lesson, following the TDTB procedure.

Input 3: Criteria for a task driven by the text (step 4 in the lesson);

Input 4: General information of different task types.

**Command 2**: Based on the provided input, design a 50-min lesson, following the TDTB approach. Choose a speaking task type that best fits B2-level EFL learners.

## 3.3 Participants

Using a convenience sampling technique, 30 English-major sophomores, including 23 males (76.67%), and 7 (23.33%) females, were selected due to their availability and willing participation in the study (Creswell & Creswell, 2023). They are English majors enrolled in a B2 Listening and Speaking Course at a Vietnamese public university in the Mekong Delta. From this group, 6 participants were subsequently selected for semi-structured interviews based on their voluntariness. The study purposes were carefully explained to all participants before they volunteered to participate.

## 3.4 Instruments and Data Analysis

### 3.4.1 Questionnaire

The TE questionnaire used to collect quantitative data consisted of two sections: (1) target participants' background information, (2) the main section which was adapted from a statistically validated questionnaire (Zare & Derakhshan, 2024) including four engagement clusters: behavioural engagement (BE), cognitive engagement (CE), emotional engagement (EE), and social engagement (SE). Each cluster consisted of 5 items and a 6-point scale (1 = strongly disagree, 2 = disagree, 3 = slightly disagree, 4 = slightly agree, 5 = agree, and 6 = strongly agree).

**Table 1** Cronbach's alpha coefficients (α) of questionnaire (n = 30)

| Dimensions | Items | α<br>Task 1 | A<br>Task 2 |
|---|---|---|---|
| BE | 1–5 | 0.828 | 0.763 |
| EE | 6–10 | 0.825 | 0.908 |
| CE | 11–15 | 0.744 | 0.846 |
| SE | 16–20 | 0.707 | 0.865 |
| Overall engagement | 1–20 | 0.846 | 0.898 |

To ensure validity, two experienced lecturers of English were involved in translating the English questionnaire into Vietnamese, the participants' mother tongue, to ensure their comprehensive understanding. The survey was then pilot tested to guarantee its content validity and internal reliability. It was later transferred to a Google Form to facilitate the assembling of data as the students were requested to respond right in class after each lesson.

SPSS v.27 was employed first to obtain the internal consistency of the dataset before Descriptive statistics, and independent-samples-t test were run. Findings from the Scale Test (Table 1) illustrate Cronbach's alpha coefficients for both overall engagement and four dimensions across two tasks were all above .07, showing a strong internal consistency.

### 3.4.2 Interviews

A semi-structured interview protocol was afterwards developed in response to four clusters of engagement and subsequently judged in relation to its content relevance and wording by two experienced researchers. The interviews were conducted online in Vietnamese after the learners had completed their questionnaires, seeking detailed qualitative proof of learners' engagement in two tasks. A total of 6 participants were randomly selected from the questionnaire phase. All of them were well informed of the interview contents and purposes and were kindly invited to participate in the interview through Google Meet. The researcher asked for participants' permission to record their oral responses (Table 2).

**Table 2** Demographics of interview participants (n = 6)

| Tasks | Students | Gender | Majors |
|---|---|---|---|
| Task 1 | A | Female | English language education (ELT) |
| | B | Male | ELT |
| | C | Male | ELT |
| | D | Male | ELT |
| Task 2 | E | Male | ELT |
| | F | Female | ELT |

**Table 3** Examples of coded themes and indicators

| Themes | Examples of indicators |
|---|---|
| BE | Stayed focused on task requirements<br>Tried to interact with partners to complete the task |
| EE | Very excited… comfortable… had no pressure |
| CE | Used ideas from the text plus my own knowledge<br>The topic was new… it encouraged us to think and explore |
| SE | Thought about possible ideas… asked peers/teacher for suggestions |

A thematic analysis was employed to generate relevant themes. First, audio recordings of the interviews were transcribed verbatim with TurboScribe. To ensure transcription accuracy, the researcher reviewed the recordings and manually corrected any omissions and spelling mistakes skipped by the software. Second, extracted data were analysed following a rigorous six-step procedure (Seyri & Ghiasvand, 2025): initially exploring data, creating codes, finding themes, rechecking themes, naming, and reporting themes. Below are examples of coded themes and their corresponding indicators (Table 3).

## 4 Results

Table 4 shows Descriptive Statistics of four-dimensional TE among the learners across two text-driven tasks generated by ChatGPT. Mean scores of learners' overall engagement across the two tasks were high, with Task 1 showing a slightly higher level (M = 5.07; *SD* = *0.4132*) than Task 2 (M = 4.90; *SD* = *0.4651*). Specifically, four dimensions of engagement across two AI-generated text-driven tasks were similarly high: BE1 (M = 5.15; *SD* = 0.5526), EE1 (M = 5.09; *SD* = 0.5866), CE1 (M = 5.00; *SD* = 0.5420), SE1 (M = 5.03; *SD* = 0.6769); BE2 (M = 4.89; *SD* = 0.5425), EE2 (M = 4.880; *SD* = 0.6800), CE2 (M = 4.920; *SD* = 0.5448), SE2 (M = 4.94; *SD* = 0.7468). These findings indicate that learners are highly engaged in two tasks, irrespective of differences in text types and topics.

The Independent-Samples T-test results (Table 5) showed no statistically significant differences across four dimensions of learners' engagement between two ChatGPT-generated text-driven tasks at an overall level ($t = 1.452$, $df = 58$, $p = 0.152$); and the cluster level: BE ($t = 1.886$, $df = 58$, $p = 0.064$), EE ($t = 1.301$, $df = 58$, $p = 0.198$); CE ($t = 0.618$, $df = 58$, $p = 0.539$), SE ($t = 0.507$, $df = 58$, $p = 0.614$). These non-significant differences were further evidenced through small effect sizes measured at an overall TE level ($d = 0.39$) and each TE dimension: BE (*d = 0.49)*, EE (*d = 0.33)*, CE (*d = 0.15)*, SE (*d = 0.13)* (Lovakov & Agadullina, 2021). In other words, both text-driven tasks generated by ChatGPT equally contributed to facilitating learners' engagement across all measured dimensions.

**Table 4** Descriptive statistics of EFL learners' engagement in AI-generated text-driven tasks

| | Engagement dimensions | Min | Max | M | SD |
|---|---|---|---|---|---|
| Task 1 (n = 30) | BE1 | 3.8 | 6.0 | 5.15 | 0.5526 |
| | EE1 | 3.8 | 6.0 | 5.09 | 0.5866 |
| | CE1 | 4.0 | 6.0 | 5.00 | 0.5420 |
| | SE1 | 3.0 | 6.0 | 5.03 | 0.6769 |
| | Overall | 4.25 | 6.0 | 5.07 | 0.4132 |
| Task 2(n = 30) | BE2 | 3.2 | 6.0 | 4.89 | 0.5425 |
| | EE2 | 2.6 | 6.0 | 4.88 | 0.6800 |
| | CE2 | 3.6 | 5.8 | 4.92 | 0.5448 |
| | SE2 | 3.0 | 6.0 | 4.94 | 0.7468 |
| | Overall | 3.55 | 5.70 | 4.90 | 0.4651 |

**Table 5** Independent-Samples T-test of four-dimensional engagement across two tasks (n = 30)

| | Tasks | Mean | SD | SEM | t | df | p | Cohen's d |
|---|---|---|---|---|---|---|---|---|
| BE | Task 1 | 5.15 | 0.55257 | 0.10088 | 1.886 | 58 | 0.064 | 0.49 |
| | Task 2 | 4.88 | 0.54249 | 0.09905 | | | | |
| EE | Task 1 | 5.09 | 0.58659 | 0.10710 | 1.301 | 58 | 0.198 | 0.33 |
| | Task 2 | 4.88 | 0.67996 | 0.12414 | | | | |
| CE | Task 1 | 5.00 | 0.54198 | 0.09895 | 0.618 | 58 | 0.539 | 0.15 |
| | Task 2 | 4.92 | 0.54482 | 0.09947 | | | | |
| SE | Task 1 | 5.03 | 0.67688 | 0.12358 | 0.507 | 58 | 0.614 | 0.13 |
| | Task 2 | 4.94 | 0.74676 | 0.13634 | | | | |
| Overall | Task 1 | 5.07 | 0.41329 | 0.07546 | 1.452 | 58 | 0.152 | 0.39 |
| | Task 2 | 4.90 | 0.46512 | 0.08492 | | | | |

The results were further supported with a qualitative data analysis from the semi-structured interviews. Table 6 displays the summary of reported themes.

***Behavioural engagement***

Most of the students admitted making attempts to complete the tasks and planning ideas during two tasks within the time frame. They further reported concentrating on speaking performance despite the noise caused by the activities and task challenges. Student A reported satisfaction with the time limit but wished for more time to practice, showing a desire to go beyond the minimum and engage more deeply.

> I found it comfortable performing the task in the allotted 6 min, but I wished more time to practice speaking English by completing the task. This allowed me to focus on the requirements and plan further ideas to complete the task effectively. (Student A, task 1).

Meanwhile, Student B maximised the time to fulfil task requirements, indicating strong commitment and proactive behaviour: "The task was beneficial for

**Table 6** Semi-structured interviews of learners' TE

| Themes | | BE | | EE | | CE | | | SE |
|---|---|---|---|---|---|---|---|---|---|
| | Learners | Task completion | Staying focused | Negative emotion (nervous, shy) | Positive emotion (excited, happy, confident, curious) | Summarise in own words | Connect with prior knowledge | Create examples | Peer/ teacher support |
| Task 1 | A | + | + | | + | + | + | | + |
| | B | + | | | + | | + | + | + |
| | C | + | + | + | + | | | | + |
| Task 2 | E | + | + | | + | | + | | + |
| | F | | + | | + | + | + | | + |
| | G | + | + | | + | | + | + | + |

my speaking, so I spent as much time as possible fulfilling its task requirements" (Task 1).

Two other students mentioned classroom noise but still stayed focused and completed the task, showing persistence despite distractions. Their ability to maintain concentration and effort in a disruptive environment reflects resilience and task-oriented behaviour. For example, Student C reported on task 1: "Given the noise from other students' speaking performances, I managed to stay focused on task content and complete it," and Student E mentioned task 2, "the class was noisy but I tried to get the task completed because it actually helped me practice my speaking skills."

***Emotional engagement***

Five out of six participants expressed their greater interest and less anxiety in performing the tasks due to familiar contents and real-world situations. For example, Student A commented, "I felt interested in performing the task whose topic and content were familiar to my experience. The task [task 1] content had gone viral on social media", and Student G shared a similar idea, "I didn't feel pressured to perform the task [task 2] in 6 min because the task content was familiar to me. It was about a famous female vendor yelling at her customers featured on TV".

Tasks that simulate authentic scenarios boosted their feelings of preparedness and confidence. Student C emphasised how task 2 improved their speaking performance: "The second task engaged me in a real-world situation, which allowed me to practice speaking and fostered my confidence during the performance". Anxiety was also reduced as students interacted and collaborated during tasks. Working together and negotiating task demands made them feel supported and less stressed. The following extract illustrates this:

> I felt less anxious when negotiating task demands and discussing ideas, which helped us manage the tasks more effectively. We were actively involved in task interaction and finally supported each other throughout task performance. (Student A, task 1)

***Cognitive engagement***

Most of the students stated that they tried to connect task contents with their prior knowledge, and to engage with the text content as input for performing tasks and language learning. As an example of cognitive effort, Student E noted how the content demands of Task 2 temptingly challenged her to brainstorm ideas for speaking: "The problem raised in the task was challenging but appealing to me because I could brainstorm to solve the problem in response to the task demands". Student G likewise admitted, "Though challenging, the task was satisfactorily completed. Each of us made great attempts to solve the problem posed in the task". This overall shows students' sustainment of cognitive engagement and commitment to task performance despite the challenge posed in the task.

In a slightly different way, Student B reported on how the text content highly engaged her while performing Task 1 because it provides her with relevant input (both ideational and linguistic), which may have facilitated her acquisition: "The text captured my own interest in reading and memorisation, which supported me in performing the task successfully because I could make use of information, vocabulary and grammar". She further stressed that the role-play activity was beneficial for her language retention by raising her awareness of how English is used in authentic situations, thereby enhancing her communication skills. In general, the text-driven tasks challenged the participants to engage in deep thinking and meaningful language learning. The cognitive load was high, but it was embraced rather than avoided.

***Social engagement***

All the participants reported their social engagement across two tasks, acknowledging the impact of interaction on their task completion irrespective of cognitive challenges in solving problems. Student B received direct support from a partner in completing task 1 in terms of "identifying correcting pronunciation and grammar mistakes during the speaking task performance". This shows active peer scaffolding—learners not only participated but took responsibility for each other's language development. Student F further reported in the interview on task 2 that "all group members participated in the idea discussions," which facilitated the task performance. Peer interaction and scaffolding also happened as the participants were actively seeking help in case of running out of ideas. For instance, Student G said, "I followed the speaking task requirements and sought help when I lacked ideas but did not lose attention to the task". This indicates that students were willing to deploy social resources for the sake of task completion and to maintain a focus on task. Such examples highlight active participation and collective cognitive involvement, which are the essentials of social engagement.

## 5 Discussion

Regarding Vietnamese EFL learners' engagement in text-driven speaking tasks generated by ChatGPT, both quantitative and qualitative results revealed positive evidence of learners' full engagement in speaking tasks driven by two different engaging text genres. These findings confirm the overall beneficial impacts of task design factors on learners' TE, echoing previous studies (e.g. Aubrey et al., 2022; Ma & Chen, 2024; Meşe et al., 2025; Nakamura et al., 2021). Together, these findings expand the current understanding that underscores the significant contributions of cognitively and affectively engaging texts as meaningful input for task design and implementation to enhance task authenticity and the assistance of ChatGPT in planning lessons to facilitate learners' TE.

Learners expressed high behavioural engagement through their efforts and sustained focus on the tasks. The texts function as input references for learners during task performance, reducing their cognitive load (Huynh & Nguyen, in press). They discussed and initiated solutions for the problems posed in text-driven speaking tasks. Meaning negotiations were actively conducted in pairs or groups in correspondence with different stages in text-driven task performances. They also made great attempts regardless of classroom noise and cognitive challenges from the problem-solving tasks. This engagement can be due to the fact that the speaking task topics were connected to their prior knowledge and were beneficial for their speaking improvement.

High enjoyment levels recorded in the study align with Nakamura et al. (2021) in that less constrained tasks were designed in response to learners' ideas and expectations. It can be inferred that tasks based on two engaging texts are somewhat aligned with learners' interest and familiarity as reflected in previous studies (i.e., Aubrey et al., 2022, Huynh & Nguyen, in press) highlighting the crucial role of familiar task topics in contributing to learners' emotional engagement (i.e., higher enjoyment). In addition, task characteristics are also seen as influential factors in all dimensions of engagement, particularly behavioural and emotional engagement, which are indicated by their active involvement in problem-solving (Task 1) and opinion-exchange (Task 2). Their emotional engagement across the tasks can also be due to the ideas prompted through the teacher–ChatGPT communication during lesson planning, highlighting Moorhouse and Wong's (2025) emphasis on the potential of GenAI, especially ChatGPT in the current study, to assist planning lessons that can enhance learners' engagement.

Learners' cognitive engagement was recorded at a high level across two tasks, reflecting the contribution of text inputs in supporting learners' task performance. Those texts address cognitive load as a potential barrier to effective task performance explained by Skehan's theory of Limited Attentional Capacities. Accordingly, learners suffer from cognitive challenges when simultaneously processing different pieces of information. This finding makes text-driven tasks particularly different from traditional ones in which task input is mainly based on the task itself. Besides, interest-based grouping and teacher support likely helped mitigate these cognitive demands, facilitating more effective engagement and task execution. Texts driving task design were cognitively achievable, highlighting comprehensible input as a facilitative condition for language learning.

Throughout the TDTB lessons, learners worked in groups and pairs based on their shared interests and friendships for text processing activities and meaning-focused negotiations. These probably accounted for the high emotional engagement recorded across the tasks, confirming Philp and Duchesne's (2016) hypothesis regarding the interaction between social engagement and other dimensions of TE. The finding also aligns with Tsoi and Aubrey (2024) that highlighted the positive effects of self-selected pairings on learner emotional engagement during task performance. High levels of social engagement observed across both tasks may be attributed to the self-selected group formation, indicating consistency with Namkung and Kim's (2024) findings in the context of computer-mediated collaborative writing, and emphasizing the positive influence of interlocutor familiarity on social engagement. Likewise, Aubrey et al. (2022) revealed that students' engagement was fostered in the case of cooperative learning i.e., group planning for tasks and clear task roles.

It is notable that learners' TE benefited from the pre-task phase which engaged learners in processing text content. A range of text processing activities such as readiness, text experiences, and personal responses to engaging texts were incorporated to prepare learners for text-driven task performance. These activities provided opportunities for learners to express their thoughts and emotions through task input processing (i.e., texts), thereby reducing cognitive load during the main task. This is in sharp contrast with previous studies which emphasised intensive cognitive processing in tasks. Authentic texts used as input might guarantee task authenticity and pedagogic task features in the classroom. However, as Peker (2024) notes, such authenticity can appear "puzzling" when applied within traditional task-based frameworks. In this study, the learners' engagement appeared to be enhanced by the authenticity of the text-driven tasks, which were based on two familiar and engaging authentic texts. The familiarity of these text contents created a supportive environment for interaction, particularly during speaking tasks, as the learners drew on the content to support their output. This, in turn, positively influenced their overall TE.

## 6 Conclusions

The current study has revealed positive impacts of potentially engaging texts as input factors in task design hypothesised to foster affective and cognitive engagement (Tomlinson, 2018, 2023). The findings have contributed to existing literature in that TE is encouraged by not only cognitive task complexity but also cognitive and affective text-based factors. Moreover, the TDTB approach offers a principled lens into how tasks are designed and sequenced in such a way that learners can process meaningful input from both engaging texts and tasks during performance, which might reduce their cognitive load as compared to a cognitive task-based perspective. In other words, TE from this perspective might hinder engagement due to the lack of input from which learners can take a reference during task performance, which requires them to make cognitive processing investments.

Regarding the role of ChatGPT, the study highlights that apart from using the tool to assist language learning via feedback, task-based activities as previous studies have focused on, ChatGPT can be an assistant to task-based materials development that not only reduces teacher workload but also enhances learning effectiveness. Teachers may design lesson plans by providing ChatGPT with input such as engaging texts, and a principled TDTB framework (Tomlinson, 2018, 2023) and adjusting its outputs to suit target learners.

Notwithstanding positive impacts of the approach, there are some limitations. First, limited and gender-biased sample size may not ensure generalizability to other EFL contexts. According to Hiver et al. (2024), engagement is highly dependent on and possibly affected by "cultures, communities, families, schools, peers, classrooms, and specific tasks and activities within those classrooms" (p. 203). This calls for extensive research regarding the approach in a wider context. Second, learners' TE was examined only within two short lessons, so further studies may follow a longer-time design to unpack how learners' engagement changes over a series of tasks. Besides, the types of tasks selected in the study may limit pedagogical implications.

Future research can employ a variety of tasks, following the TDA with the support of ChatGPT. Finally, the TDTB lessons were conducted in a classroom context which might involve the impacts of extraneous variables such as noise and uneven English language proficiency among learners. Research into such factors may leverage our understanding of a technology-assisted TDA.

**Acknowledgements** This study used ChatGPT to generate two task-based lessons based on two selected texts. Thus, the lesson samples are not included in this chapter.

**Conflict of Interest** The authors have no conflict regarding this publication.

**Ethics** The participants consented to this study. This manuscript is original. It has not been published anywhere or is under consideration for publication.

## Appendices

## Text sources

Text source for lesson 1: https://saigoneer.com/saigon-food-culture/8237-hanoi-s-hot-tempered-b%C3%BAn-ch%E1%BB%ADi-vendor-featured-on-anthony-bourdain-s-cnn-show
Text source for lesson 2: https://www.youtube.com/watch?v=7646b5Ee7BE

## References

Aoyama, T., Yamazaki, J. S., Nakamura, S., Vuogan, A., An, H., Kim, C. J., & Al-Hoorie, A. H. (2024). Conceptualization and operationalization in L2 task engagement research: Taking stock and moving forward. *Language Teaching, 57*(4), 797–601. https://doi.org/10.1017/S026144482400020X

Aubrey, S., King, J., & Almukhaild, H. (2022). Language learner engagement during speaking tasks: A longitudinal study. *RELC Journal, 53*(3), 519–533. https://doi.org/10.1177/0033688220945418

Creswell, J. W., & Creswell, J. D. (2023). *Research design: Qualitative, quantitative, and mixed methods approaches* (6th ed.). SAGE Publications.

Creswell, J. W., & Guetterman, T. C. (2024). *Educational research: Planning, conducting, and evaluating quantitative and qualitative research* (7th ed.). Pearson.

Dao, P. (2021). Effects of task goal orientation on learner engagement in task performance. *International Review of Applied Linguistics in Language Teaching, 59*(3), 315–334. https://doi.org/10.1515/iral-2018-0188

Ellis, R. (2009). The differential effects of three types of task planning on the fluency, complexity, and accuracy in L2 oral production. *Applied Linguistics, 30*(4), 474–509. https://doi.org/10.1093/applin/amp042

Ellis, R. (2024). Conclusion: Some thoughts on investigating individual differences in task-based language teaching. In S. Li. (Ed.), *Individual differences and task-based language teaching* (pp. 346–364). John Benjamins.

Fathi, J., Rahimi, M., & Derakhshan, A. (2024). Improving EFL learners' speaking skills and willingness to communicate via artificial intelligence-mediated interactions. *System, 121*, Article 103254. https://doi.org/10.1016/j.system.2024.103254

Fredricks, J. A., Blumenfeld, P. C., & Paris, A. H. (2004). School engagement: Potential of the concept, state of the evidence. *Review of Educational Research, 74*(1), 59–109. https://doi.org/10.3102/00346543074001059

Hiver, P., & Wu, J. (2023). Engagement in TBLT. In C. Lambert, S. Aubrey, & G. Bui (Eds.), *The role of the learner in task-based language teaching* (pp. 74–90). Routledge. https://doi.org/10.4324/9781003227267

Hiver, P., Al-Hoorie, A. H., Vitta, J. P., & Wu, J. (2024). Engagement in language learning: A systematic review of 20 years of research methods and definitions. *Language Teaching Research, 28*(1), 201–230. https://doi.org/10.1177/13621688211001289

Huynh, S. T., & Nguyen, L. V. (in press). Exploring learners' emotional engagement in text-driven task-based lessons: Implications from an EFL setting. *Language Related Research.*

Karataş, F., Abedi, F. Y., Ozek Gunyel, F., Karadeniz, D., & Kuzgun, Y. (2024). Incorporating AI in foreign language education: An investigation into ChatGPT's effect on foreign language learners. *Education and Information Technologies*, 1–24. https://doi.org/10.1007/s10639-024-12574-6

Lai, C., & Sundqvist, P. (2025). Research agenda: Synergizing in-class and out-of-class language learning with technology. *Language Teaching,* 1–22. https://doi.org/10.1017/S026144482500014X

Lambert, C. (2017). Tasks, affect and second language performance. *Language Teaching Research, 21*(6), 657–664. https://doi.org/10.1177/1362168817736644

Liu, S., & Yu, G. (2022). L2 learners' engagement with automated feedback: An eye-tracking study. *Language Learning & Technology*, *26*(2), 78–105. https://doi.org/10.64512/10125/73480

Lo, C. K., Hew, K. F., & Jong, M. S. Y. (2024). The influence of ChatGPT on student engagement: A systematic review and future research agenda. *Computers & Education*, 105100. https://doi.org/10.1016/j.compedu.2024.105100

Lovakov, A., & Agadullina, E. R. (2021). Empirically derived guidelines for effect size interpretation in social psychology. *European Journal of Social Psychology, 51*(3), 485–504. https://doi.org/10.1002/ejsp.2752

Ma, Y., & Chen, M. (2024). AI-empowered applications effects on EFL learners' engagement in the classroom and academic procrastination. *BMC Psychology, 12*(1), 739. https://doi.org/10.1186/s40359-024-02248-w

Meşe, E., Kuru, Ö., & Atay, D. (2025). Facilitating student engagement through ChatGPT. *The Journal of Educational Research*, 1–11. https://doi.org/10.1080/00220671.2025.2510394

Moorhouse, B. L., & Wong, K. M. (2025). Generative Artificial Intelligence and language teaching. *Cambridge University Press.* https://doi.org/10.1017/9781009618823

Muniandy, J., & Selvanathan, M. (2025). ChatGPT, a partnering tool to improve ESL learners' speaking skills: Case study in a Public University, Malaysia. *Teaching Public Administration, 43*(1), 4–20. https://doi.org/10.1177/01447394241230152

Nakamura, S., Phung, L., & Reinders, H. (2021). The effect of learner choice on L2 task engagement. *Studies in Second Language Acquisition, 43*(2), 428–441. https://doi.org/10.1017/S027226312000042X

Namkung, Y., & Kim, Y. (2024). Learner engagement in collaborative writing: The effects of SCMC mode, interlocutor familiarity, L2 proficiency, and task repetition. *System, 121*, Article 103251. https://doi.org/10.1016/j.system.2024.103251

Nguyen H, H. (2025). AI for English speaking practice: A study of effectiveness and engagement among Vietnamese university learners. *Australasian Journal of Educational Technology, 41*(3), 45–61. https://doi.org/10.14742/ajet.10050

Pan, M., Guo, K., & Lai, C. (2024). Using Artificial Intelligence chatbots to support English-as-a-foreign language students' self-regulated reading. *RELC Journal, 0*(0). https://doi.org/10.1177/00336882241264030

Peker, B. G. (2024). Task authenticity and materials design. In B. G. Peker & A. Acar (Eds.), *Developing and designing materials for English language teaching and learning* (pp. 1–20). Cambridge Scholars Publishing.

Philp, J., & Duchesne, S. (2016). Exploring engagement in tasks in the language classroom. *Annual Review of Applied Linguistics, 36*, 50–72. https://doi.org/10.1017/S0267190515000094

Richards, J. C. (2022). Exploring emotions in language teaching. *RELC Journal, 53*(1), 225–239. https://doi.org/10.1177/0033688220927531

Seyri, H., & Ghiasvand, F. (2025). Emotions at the heart of tensions: A transcendental phenomenology of Iranian EFL teachers' intra-and interpersonal emotional tensions and regulation strategies. *International Journal of Applied Linguistics, 35*(2), 768–781. https://doi.org/10.1111/ijal.12655

Singha, S., Singha, R., & Jasmine, E. (2024). Enhancing language teaching materials through artificial intelligence: Opportunities and challenges. In F. Pan (Ed.), *AI in language teaching, learning, and assessment* (pp. 22–42). IGI Global. https://doi.org/10.4018/979-8-3693-0872-1.ch002

Songsiengchai, S. (2025). Implementation of Artificial Intelligence (AI): Chat-GPT for effective English language learning among Thai students in higher education. *International Journal of Education and Literacy Studies, 13*(1), 302–312. https://doi.org/10.7575/aiac.ijels.v.13n.1p.302

Svalberg, A. M. L. (2018). Researching language engagement; current trends and future directions. *Language Awareness, 27*(1–2), 21–39. https://doi.org/10.1080/09658416.2017.1406490

Teng, M. F. (2024). "ChatGPT is the companion, not enemies": EFL learners' perceptions and experiences in using ChatGPT for feedback in writing. *Computers and Education: Artificial Intelligence, 7*, Article 100270. https://doi.org/10.1016/j.caeai.2024.100270

Tomlinson, B. (2013). Second language acquisition and materials development. In B. Tomlinson (Ed.), *Applied linguistics and materials development* (pp. 11–29). Bloomsbury.

Tomlinson, B. (2018). Text-driven approaches to task-based language teaching. *Folio, 18*(2), 04–07.

Tomlinson, B. (2023). Using text-driven and other principled frameworks for developing language learning materials. In T. Brian (Ed.), *Developing materials for language teaching* (pp. 99–127). Bloomsbury Publishing.

Tran, N. T., & Nguyen, T. T (2024). Revitalizing English language learning: an exploration of ChatGPT's impact on student engagement and motivation. In H. P. Bui, R. Kumar, & N. K. Kamila (Eds.), *Innovations and applications of technology in language education* (pp. 98–112). Auerbach Publications. https://doi.org/10.1201/9781003473916

Tsoi, S. C., & Aubrey, S. (2024). Impact of student-selected pairing on collaborative task engagement. *ELT Journal, 78*(1), 42–53. https://doi.org/10.1093/elt/ccad024

Xin, Z., & Derakhshan, A. (2025). From excitement to anxiety: Exploring English as a foreign language learners' emotional experience in the artificial intelligence-powered classrooms. *European Journal of Education, 60*(1), Article e12845. https://doi.org/10.1111/ejed.12845

Yuan, L., & Liu, X. (2025). The effect of artificial intelligence tools on EFL learners' engagement, enjoyment, and motivation. *Computers in Human Behavior, 162*, Article 108474. https://doi.org/10.1016/j.chb.2024.108474

Zare, J., & Derakhshan, A. (2024). Task engagement in second language acquisition: a questionnaire development and validation study. *Journal of Multilingual and Multicultural Development*, 1–17. https://doi.org/10.1080/01434632.2024.2306166

Zhang, Y., & Miao, Z. (2025). Enhancing EFL learners' engagement and motivation through immersive technologies: The role of artificial intelligence, augmented reality, virtual reality, and mobile applications. *European Journal of Education, 60*(2), Article e70128. https://doi.org/10.1111/ejed.12750

Zhou, C., & Hou, F. (2024). Can AI empower L2 education? Exploring its influence on the behavioural, cognitive and emotional engagement of EFL teachers and language learners. *European Journal of Education, 59*(4), Article e12750. https://doi.org/10.1111/ejed.12750

**Huynh Truong Sang** is currently a PhD candidate in TESOL methodology, working as an early-career teacher educator at An Giang University, Vietnam National University Ho Chi Minh City. His research interests involve English teaching methodology and second language acquisition.

**Nguyen Van Loi** is currently Associate Professor in ELT theory at Can Tho University. He has been engaged extensively in English teacher education and professional development for high school teachers and university instructors in the Mekong Delta. His main areas of research involve English teaching methodology, teacher cognition and teacher development in ELT.

# Artificial Intelligence-Supported Willingness to Communicate and Intercultural Communication

**Thanh Tien Nguyen and Hung Phu Bui**

**Abstract** The evolution of technology has provided opportunities for out-of-class interaction in L2 English contexts, where exposure to English is restricted. L2 learners can interact with artificial intelligence (AI) to achieve expected objectives and goals. Also, AI-mediated platforms have been developed to facilitate learners' intercultural communication. Considering the potential of human-AI interaction, this chapter argues that AI-mediated platforms can be used to improve intercultural communication and willingness to communicate. The chapter first reviews the literature on AI and then discusses its potential to improve L2 learners' intercultural communication and willingness to communicate. Regarding the current shortcomings and ongoing evolution of AI, we finally argue that AI will develop further to assist L2 in interaction, especially with those from other cultures.

**Keywords** Artificial intelligence · Intercultural communication · Willingness to communicate

## 1 Introduction

Developing the competence to communicate effectively in English has been proven an ultimate goal of language education (Liddicoat & Scarino, 2013; Saito & Hanzawa, 2015). This has paved the way for growing attention to a concept called willingness to communicate (WTC), putting one's language competence into practice. However, achieving this goal seems troublesome as some learners, even at an advanced proficiency level, may hesitate be involved in communication (Dörnyei, 2005). The problem may be even further exacerbated due to the rise of globalization, in which

T. T. Nguyen (✉)
Hoa Sen University, Ho Chi Minh City, Vietnam
e-mail: thanh.nt01199@sinhvien.hoasen.edu.vn

H. P. Bui
Faculty of Foreign Languages, Ton Duc Thang University, Ho Chi Minh City, Vietnam
e-mail: buiphuhung@tdtu.edu.vn



H. P. Bui and L. P. H. Huong (eds.), *Artificial Intelligence in Education and Academic Research*, Studies in Computational Intelligence 1269,
https://doi.org/10.1007/978-3-032-18392-7_8

learners of English as a second or foreign language (L2) are required to communicate with people from cultures other than their own. This may, according to the contemporary literature be attributed to such factors as intercultural competence (ICC), referred to as one's competencies and skills to communicate with those from different cultures, that may be taken into account to ensure intercultural interaction is successfully accomplished (Byram, 2012).

The evolution of artificial intelligence (AI) has transformed the landscape of education. According to Chen et al., and and's (2020a, 2020b) comprehensive review of articles published in international high-quality journals from 2009 to 2020, AI was widely adopted and utilized across various educational domains, including administration, instruction, and learning processes. Such endeavours demonstrated the potential of AI in mediating communication in the digital learning modality. However, there exists little literature on how to effectively leverage AI to enhance L2 English learners' WTC.

Given that WTC can be influenced by the interlocutor, setting, and ICC, we would argue for the transfer of learners' WTC with AI tools to promote learners' desire or readiness to be involved in human communication in the intercultural discourse. To contribute to the current literature, the current chapter provides a critical view on utilizing AI to mediate intercultural communication to promote their willingness to communicate. Initially, it will briefly outline the factors that may impact one's WTC and ICC to explain the challenges for intercultural communication in L2 contexts. Subsequently, how to employ AI-mediated platforms to promote WTC between those of different cultures will be critically reviewed.

## 2 Willingness to Communicate

L2 WTC, broadly referring to "readiness to enter into the discourse at a particular time with a specific person or persons by using L2" (MacIntyre et al., 1998), may be influenced by many factors (mainly linguistic, psychological, and social) that can be directly or indirectly associated with this construct (Dörnyei, 2005). According to Bui et al. (2022) these factors can be categorized into internal or external factors, including student-related, instructor-related, task-related, and setting-related. For instance, MacIntyre and his colleagues (1998) argue that two easy-to-identify internal determinants of WTC are communication fear and perceived communication ability. Accordingly, students' self-perceived communication ability may influence their WTC. If students are not confident enough, they may choose not to participate in communication. For example, students may hesitate to communicate if they think they can be criticized for their errors in pronunciation, grammar, or content, so they may lose their confidence and remain silent in communication (Bui et al., 2022). Anxiety also plays a role as it results in negative feelings such as self-doubt, or shyness when interacting with other people (Zhang et al., 2025).

According to Hoang and Bui (2023), a low-anxiety environment may encourage students to take part in conversation. They feel more confident if they perceive the environment as safe enough for them to express their ideas and thoughts. Moreover, a higher WTC level is also witnessed when students communicate with familiar interlocutors, who they have gotten to know before. However, it is necessary to note that, in educational settings, students may refuse to communicate if they think the environment has potential threats to their face or the teacher or peers are not willing to listen to them and/or support their communication.

In addition, students with opportunities of exposure to the target language may develop a more positive attitude to the language, thereby resulting in their improving WTC (Alemi et al., 2013). This might be because students who have engaged in conversations with others might be more aware of cultural differences and develop a more positive attitude toward using the L2.

Another factor may be related to culture. Sato's (1990) study demonstrated that although students coming from Asia accounted for about two-thirds of her university class, they only took approximately one-third of the speaking turns. This may come from the concept of "face" which varies across cultures and explained by the study of Yates and Trang (2012). Vietnamese students learning in Australia taking part in the study often described themselves as "passive", "quiet" or "not as active as local students". Female students, in particular, attributed this behavior to Vietnamese cultural values. These included the expectation that juniors should obey seniors, the emphasis on modesty and restraint in speech, the desire to avoid standing out in formal settings, and a preference for indirect communication with figures of authority.

The learning tradition also contributes to this unwanted silence. The learning culture of Vietnam is often viewed as passive, teacher-centred, or exam-oriented. The teacher is usually considered knowledge providers, and students are expected to listen and make notes more than to raise their voice. Moreover, Vietnamese students may also think that if there is nothing valuable to contribute, they should remain silent (Yates & Trang, 2012).

## 3 Intercultural Communication

WTC might become more challenging when it comes to talking with other people of different cultural backgrounds. As Liddicoat and Scarino (2013) put it, culture is "a complex system of concepts, attitudes, values, beliefs, conventions, behaviors, practices, rituals, and lifestyle of the people who make up a cultural group, as well as the artifacts they produce and the institutions they create". In addition, culture is dynamic and changeable, which is nearly similar to the traits of a language. Therefore, communication is not just a combination of grammar and vocabulary, but it is also a cultural factor as nearly all messages being communicated are through a cultural context. When people speak another language, they do not just only exploit the functions of the language, but they also start to function within that culture (Hamadouch & Aoumeur, 2024; Liddicoat & Scarino, 2013). Native speakers might be tolerant of grammatical or vocabulary errors, but cultural misunderstandings can cause serious miscommunication. This is largely because people might have less awareness of the cultural rules that govern interaction compared to grammatical structures or lexical items. Therefore, communicative competence is not really the objective that language learners should aim for. Instead, in bicultural and multicultural settings, learners may need ICC in order to be a successful language user (Byram, 2012).

Developing intercultural knowledge may require learners to tolerate cultural differences. In language learning, two approaches are acknowledged to be effective: decentering from one's culture when talking with others (Liddicoat & Scarino, 2013). Unfortunately, cultural knowledge, including cultural traits that are usually unnoticed unless the learners perceive their importance, is not something the learner can easily pick up by themselves. Instead, it is acquired from real-life communication. In the L2 English classroom, the instructor is crucial as they are supposed to assist students in developing communication competence. However, according to Chau and Truong's (2018) research, in the context of Vietnamese classrooms, Vietnamese teachers did not often integrate intercultural teaching in the classroom practice to enhance students' ICC although they fully understood the importance and practicality of intercultural teaching. In a similar vein, Nguyen's (2013a, 2013b) study found that Vietnamese teachers often reported that their heavy workload, such as the demand to teach the language elements for assessment, time constraints, teaching materials, and large class size, prevented them from teaching the culture.

Additionally, students may not know the role of culture in language education. According to Nguyen's (2013a, 2013b) survey conducted in Vietnam, students often focused more on their language development and neglect cultural factors. Some may also consider passing the test or exam as their main objective of learning. If there are no examinations that require cultural knowledge, very little attention is paid to culture. Thus, those students do not have the motivation to explore cultural similarities and differences.

A lack of exposure to foreign culture is also an issue. Students living in EFL countries, including Vietnam, do not have many chances to interact with English native speakers as English is not widely spoken in their everyday life. Although textbook authors have tried to deal with this by dedicating specific sections for cultural topics, those books often overlook cultural information that could be important for learners when interacting with native speakers.

The teacher is also responsible for students' intercultural incompetence. This is because teachers themselves may have limited cultural knowledge or cultural information is rarely discussed in the classroom (Munandar, 2019; Novita & Purwati, 2021). Most teachers gain cultural knowledge from books or self-study, thus having little chance to engage with other cultures. Therefore, many of them have not developed sufficient intercultural skills and awareness, so may find it difficult to transmit this knowledge to their students (Nguyen, 2013a, 2013b).

## 4 The Relationship Between Willingness to Communicate and Intercultural Communication

As in the aforesaid, improving both WTC and ICC is fundamental in EFL teaching & learning. The absence of any element can result in difficulty in interacting with foreigners. By contrast, having both elements can result in effective intercultural communication. In fact, there is a close relationship between those traits. From the study conducted with 400 12 Filipino graders, Juhra et al. (2024) conclude that learners who are aware of cultural differences and similarities, open-minded to new values and traditions, and skilled in managing cross-cultural communication, have a higher tendency to communicate effectively.

ICC is acknowledged as a key determinant of WTC. Miauw and Guo's (2021) research highlighted the strong link between intercultural competence and learners' WTC, stressing that if the learners can be adaptable and flexible towards the cultures, they can enhance their WTC. The researchers emphasize that when speakers may feel comfortable with the culture of the language they learn, they tend to use it in communication without effort. Similarly, in the schematic representation of WTC proposed by Dörnyei (2005), ICC also appears. In particular, Layer IV, which connects WTC to motivational properties such as motivation between individuals and between ethnic groups and self-confidence, propose that one's positive attitudes toward a particular language and its speakers may significantly boost WTC.

## 5 Leveraging AI to Improve Willingness to Communicate and Intercultural Communication

Research has consistently shown AI to substantially improve various aspects of English-speaking competence (Fathi et al., 2024). AI has attracted attention of researchers and practitioners with interests in AI-mediated WTC and ICC. For instance, Rahimi and Fathi's (2024) experimental study showed that AI-assisted learning could promote L2 English students' willing to speak than traditional face-to-face learning. Similarly, students' ICC can be improved through AI use, as confirmed by Klimova and Chen (2024), who systematically reviewed 11 empirical studies conducted globally. In general, AI has made a profound impact on students' language and cultural learning.

AI may have impact on learners' L2 WTC during their language learning processes. The interactions generated by using AI-mediated applications can result in improvements in language proficiency. This is because AI may play the role of an advanced speaker, who provides comprehensible input for users to refine their language output by negotiating meanings or requesting clarifications. This interactive process allows leaners to modify and adjust their language output, eventually improving their language competence (Gass, 2003; Long, 1983). Although language proficiency is acknowledged as one factor in determining students' WTC, a generally positive association between actual and perceived proficiency has been recorded by previous researchers (Al-Murtadha, 2021; Menezes & Juan-Garau, 2015; Takahashi, 2009). That is, AI-mediated communication shows its potential to foster communication competence.

Communication anxiety is also documented in the current literature to significantly hold students back from engaging in L2 conversations (MacIntyre et al., 1998). Fortunately, AI-powered platforms offer a solution to successfully tackle this issue. Conversational AI applications, such as Elsa or Duolingo, provide opportunities for a low-stress speaking environment, which may gradually build learners' confidence (Tai, 2022). This occurs because the influence of the Eastern learning tradition and face culture is minimized. Instead of interacting with their peers or teachers, learners mostly interact with an anonymous chatbot, who does not judge or criticize their mistakes (Tai & Chen, 2020). Without the embarrassment commonly perceived when making errors in physical classroom settings (Bui & Nguyen, 2022), students may be more ready to participate in communicating in the target language (Jeon et al., 2023; Lee & Hsieh, 2019; Zou et al., 2024).

The limited access or even absence of intercultural interactions can also result in challenges in language learning, especially in countries like Vietnam, where the target language is not used frequently. AI-powered tools, which may be used as virtual language peers or culturally adaptive chatbots, show their potential to bridge this gap by providing opportunities of exposure to authentic language in various communication scenarios in different cultural contexts. By having more chances to interact with cultural elements, even if the interactions are not face-to-face, students can gain a deeper understanding of the diversity of cultures, thus improving both their communications skills and cultural awareness (McCallum, 2004). As learning can be co-constructed through interaction (Vygotsky, 1978), online interaction proves useful for improving students' language and ICC, followed by a correlative improvement in their L2 WTC (Miauw & Guo, 2021).

Students' confidence in using the target language to talk with real native speakers can be boosted thanks to language translation, a notable feature of AI technology. Tools like Google Translate provide real-time and comprehensible translation, enabling people with different cultural backgrounds to collaborate regardless of their origins (Ghafar et al., 2023; Li & Zhang, 2022; Mohamed et al., 2024). Moreover, the ability to consider the whole sentence instead of one particular word (Vaswani et al., 2017) and handle multiple languages at a time (Conneau et al., 2020; Pires et al., 2019) allows these tools to reduce misunderstanding and contribute to better comprehension. This suggests a higher level of confidence may be gained from frequent intercultural interactions with the help of translation tools, potentially transferring to a higher level of L2 WTC.

Finally, AI can speed up the process of developing ICC. Fully adapting to a culture is possibly a lengthy and complicated process, requiring the learners to experience different stages. According to the model of intercultural sensitivity (Bennet, 1986), there are six stages a learner must go through to learn about a culture, from ethnocentric (denial, defense, minimization) to ethnorelative stages (acceptance, adaptation, integration). AI has the potential to accelerate and simplify this progression through simulations or virtual exchanges. For example, a survey conducted by Sarwari et al. (2024) showed AI tools with virtual reality can place students in a cultural context that they are learning, thereby making abstract cultural concepts more tangible and approachable.

Despite the effectiveness of AI, it unfortunately presents several challenges. Firstly, although efforts have been made to finetune AI models on culturally sensitive datasets (Liu et al., 2021), understanding subtle cultural differences and nuances remains a significant challenge that AI-powered tools are not currently capable of (Huang et al., 2020). Cross-cultural communication involves many aspects and variations, such as behaviors, beliefs or practices, which have not been yet interpreted successfully by AI technology. Ismailov (2021) and Khasawneh (2023) explained that because of these limits, AI can easily cause confusion or misunderstandings, especially in complex conversations. Secondly, Ou and Malmström (2023) and Sarwari et al. (2024) warned that cultural bias in AI systems can distort intercultural meanings. If a system learns from data that is bias, it is not likely to properly respect different cultures. The concern is that platform will just copy and spread the unfair information it was taught (Binns et al., 2018; Klimova & Chen, 2024).

It may be essential that the teacher facilitates students' communication by engaging them in communicative tasks, which can highly impact their WTC (Cao & Philp, 2006; MacIntyre & Legatto, 2011). Language teachers must be aware of the difficulties that their students are facing. As discussed above, there are a variety of factors that can be a barrier to WTC and ICC. Language incompetence might be a challenge to successful communication, but it is not the only reason why students refuse to engage in communication (MacIntyre et al., 1998), especially with people from other cultures (Henry & MacIntyre, 2024). As there is a clear connection between WTC and ICC (Miauw & Guo, 2012), they should be seen as a common goal that should be achieved together, not separately (Gao, 2006). The insights from the studies above suggest several teaching implications for improving both intercultural competence and WTC among Vietnamese students.

Considering the face-saving culture that has been mentioned before, the role of a teacher is not to eliminate this rooted tradition, but to adjust their lessons to minimize its impact. It may be necessary for language instructors to be attentive to the sociocultural factors that affect WTC and adopt suitable teaching methods to cope with them. Firstly, teachers must try to create a safe environment to reduce communication anxiety and promote students' confidence. One method is to facilitate peer collaboration through which students can cooperate or collaborate to achieve a shared goal (Philp, 2013). Students can have some time to get to know their classmates before working together. When they are familiar with each other, it is easier for them to share their personal stories and thoughts (Mystkowska Wiertelak & Pawlak, 2017; Wei & Cao, 2021). This should be always done even when the students live in the same community. Those coming from the same area may represent diverse backgrounds, with different experiences and mindsets. These differences could be perceived as "information gap", which in turn foster curiosity and authentic communication needs (Gallagher & Robins, 2015). Moreover, different viewpoints on the same issue can be beneficial in intercultural exchanges, which may then become a source to incorporate into the lesson (Hoa, 2020).

A low-stress speaking environment can also be developed from the assistance of AI. Teachers might use AI chatbots as virtual interlocutors to simulate authentic conversations. For example, teachers can use ChatGPT or Google Assistant, which has been rendered as an 'ideal interlocutor' to whom L2 learners may be more willing to talk with (Peng & Liang, 2025; Tai & Chen, 2020), to support their teaching. This application demonstrates noticeable abilities that enable human-like dialogue, including answering questions, providing word definitions, and the capacity to "talk back", such as admitting errors or building conversations upon users' previous responses (Barrot, 2023; Kohnke et al., 2023). Most importantly, learners can escape from the fear of being judged when interacting with AI, thereby becoming more willing to speak.

Secondly, it is possible that teachers be aware that female students may avoid speaking in public and raising their voices. Teachers may explain their expectations to the students, be open-minded to new ideas and opinions, and avoid being judgmental, especially when the student's answers may be incorrect (Peng, 2014). The key here is to encourage students to engage by their own will rather than forcing them to do so (Bui et al., 2022).

In this sense, AI, with its various interesting features, can mediate students' language learning, suggesting students' satisfaction from interaction with AI tools. For instance, interview data from Zou et al.'s (2025) research revealed that the interactive features and comprehensive functions of EAP Talk could attract students, and most of the participants in the study showed their willingness to use the platform in the future. Therefore, instead of limiting or even prohibiting, teachers should encourage students to use AI to practice speaking if it is their interest (Peng & Wu, 2022; Zhang et al., 2024). Similar to the classroom environment, positive attitudes toward AI can strongly facilitate beyond-classroom learning (Liu & Ma, 2023).

In addition, students, especially those with a higher level of language competence, should be equipped with basic scaffolding techniques to help their peers during group work. The concept of scaffolding derives from Vygotsky's (1978) sociocultural theory, which states that it could occur between peers who are at similar stages of development. Peer scaffolding can also enable students to be exposed to diverse expertise and opinions (Nguyen, 2013a, 2013b), which are a crucial element in building their ICC. However, it is not easy to instruct students, who do not receive professional teaching training, to scaffold each other. Teachers must therefore pair groups strategically, e.g. grouping higher and lower students together (Le, 2006), and assign roles to allow students to support each other during group work. Also, teachers may model some scaffolding techniques such as asking questions like "Could you explain …?" or "What if we try …?" to encourage their students to apply those during their discussion. Moreover, scaffolding can be implemented within virtual learning environments. In some intelligent tutoring systems, adaptive scaffolding delivers in many forms, including pop-up messages, advice from a pedagogical agent, or conversation with an automated agent (Graesser, 2016). Therefore, along with teaching students in-class scaffolding techniques, teachers should also demonstrate how to scaffold AI, such as how to generate constructive prompts to break down complex tasks into smaller steps.

Building a good student–teacher rapport could also be an effective way. Talking about topics that are relevant to students or being humorous with funny stories might help generate students' interest in the lesson and the teacher themselves. This could help reduce anxiety and create an environment that encourages "low filter", which could make input more comprehensible according to the Affective Filter Hypothesis proposed by Stephen Krashen (1982). This relationship plays an even more important role currently regardless of the impact of the technologization of education. In other words, while making use of technology, teachers must also recognize and preserve the vital relationships between them and students. A simple appearance of dialogue generated by AI cannot replace real human relations, which are built on understanding, empathy or emotional connection between individuals. The aim of education is not only developing a particular skill, but also shaping a person's characteristic (Guilherme, 2019). Moreover, a strong bond developed between the instructors and students can reduce the negative effect of the "power distance" commonly found in Eastern classrooms and encourage students to speak more (Peng, 2014).

Furthermore, teachers should create communicative tasks that highlight the importance of cultural interactions. Students are more likely to engage in conversations if the topics being discussed are cognitively or experientially familiar to them (Mystkowska-Wiertelak & Pawlak, 2017; Peng, 2012). As an expert, the teacher should try to create the connections between the topics and the students, linking the cultural elements in the topic with the student's pre-existing knowledge. The teacher should encourage noticing, comparing, and making intertextual and experiential connections while prompting reflection on language, culture, communication, and learning in diverse contexts. For example, they should listen actively to their learners, consider every response, and come up with some interactive questions to foster these connections to foster students' WTC (Hejazi et al., 2023; Khajavy et al., 2016; Peng & Woodrow, 2010). Additionally, teachers can use AI tools inside the classroom to simulate conversational scenarios that students might encounter in a different cultural context to experience different communication strategies and receive instant feedback. Combining AI chatbots or learning platforms with traditional or inquiry-based activities increased students' cultural awareness, and communication skills (Ismailov, 2021).

Language learning should not be limited to linguistic elements but should also include cultural dimensions. Therefore, teaching resources should not only be limited to textbooks although they are still a great source of cultural teaching. Textbooks may be supplemented by authentic materials from a range of sources: written texts, video or audio texts, music, multimedia, and so on. Clearly, AI technologies, whether used for informational or social purposes, have the potential to enhance intercultural learning so teachers must learn to make use of them effectively. Hoa (2020) also suggests that the application of the IcLLT (Intercultural language learning & teaching) model with adapted intercultural language activities can help improve the situation. The 5-stage model, proposed by Liddicoat and Scarino (2013), is an effective way of enhancing student's ICC and has been proven to be applicable in the context of Vietnamese classrooms with slight modifications (Hoa, 2020).

Assessment practices should reflect the importance of both IC and WTC. Assessment should pay more attention to WTC by rewarding active participation and risk-taking in communication, rather than focusing solely on language elements. Teachers should also provide formative feedback on students' intercultural and communicative efforts. This can be done more effectively with the use of AI, which can provide support for interpreting non-verbal gestures and instruct students to use the language culturally appropriately (Dong, 2023). Similarly, the study by Dizon (2017) showed that indirect pronunciation feedback found in Alexa, an AI-powered assistant, improved students' effectiveness in storytelling, paving the way for more authentic conversations. This might help students develop intercultural awareness and encourage them to reflect on cultural experiences.

Finally, since AI can spread misinformation or misinterpret cultural nuances, teachers must show students how to critically use AI. Students need to know how to use AI appropriately, especially in discriminating between ethical and unethical uses (Robinson, 2020; Rodríguez-García et al., 2020). The quest for AI literacy has recently received lots of attention, from introducing AI concepts to children (Gong et al., 2020; Wong et al., 2020) to further developing the AI abstract knowledge for adult learners (Kandlhofer et al., 2016). Students, regardless of their age, need to be taught how to know, understand, use, and especially evaluate AI (Ng et al., 2021). Being critical in utilizing AI can foster both intercultural awareness and willingness to take part in cross-cultural conversations. To equip students with this capacity, teachers must keep themselves up to date with the latest AI concepts, especially the ones that are potentially relevant to their students' learning. Then, they can select appropriate teaching methods based on their learners' characteristics to deliver IT-related knowledge in the most suitable way (Ng et al., 2021).

## 6 Conclusion

WTC and ICC show their vital roles in L2 learning. Students at a high proficiency level may acquire the language more effectively due to frequent AI-supported exposure to authentic language of which ICC is an integral component. However, it is to note that some factors may negatively influence students' WTC and ICC, making it difficult for students to be involved in discourse with people from other cultures. These can come from the students themselves (lack of confidence/perceived language competence and high levels of anxiety), the teachers (lack of cultural knowledge or awareness of intercultural learning), or external factors (textbooks focusing on linguistic elements only, differences in learning cultures or little chance of cross-cultural interaction). With the development of AI technologies, some of these factors can be effectively dealt with. AI can increase student's perception of language proficiency, reduce their anxiety, create more opportunities for intercultural interactions. To improve students' confidence and ability to communicate with people from other cultures, AI proves to be a recommended option.

**Acknowledgements** The first author used ChatGPT to proofread this manuscript.

**Conflict of Interest** The authors have no conflict regarding this publication.

**Ethics** This manuscript is original. It has not been published anywhere or is under consideration for publication.

## References

Alemi, M., Tajeddin, Z., & Mesbah, Z. (2013). Willingness to communicate in L2 English: Impact of learner variables. *Journal of Research in Applied Linguistics, 4*, 42–61. https://rals.scu.ac.ir/article_10469.html

Al-Murtadha, M. A. (2021). The relationships among self-reported and observed first language and second language willingness to communicate and academic achievement. *Language Culture and Curriculum, 34*(1), 80–94. https://doi.org/10.1080/07908318.2020.1727495

Barrot, J. S. (2023). Using ChatGPT for second language writing: Pitfalls and potentials. *Assessing Writing, 57*, Article 100745. https://doi.org/10.1016/j.asw.2023.100745

Bennett, M. J. (1986). A developmental approach to training for intercultural sensitivity. *International Journal of Intercultural Relations, 10*(2), 179–196. https://doi.org/10.1016/0147-1767(86)90005-2

Binns, R., Veale, T., & Mitchell, M. (2018). The ethical implications of biases in machine translation. *Proceedings of the 2018 Conference on Fairness, Accountability, and Transparency (FAccT 2018).* https://doi.org/10.1145/3287560.3287594

Bui, H. P., Hoang, V. Q., & Nguyen, N. H. (2022). Encouraging Vietnamese students' willingness to communicate inside English L2 classrooms. *Language Related Research, 13*(5), 1–23. https://doi.org/10.52547/LRR.13.5.17

Bui, H. P., & Nguyen, T. T. T. (2022). Classroom assessment and learning motivation: Insights from secondary school EFL classrooms. *International Review of Applied Linguistics in Language Teaching, 62*(2), 275–300. https://doi.org/10.1515/iral-2022-0020

Byram, M. (2012). *From foreign language education to education for intercultural citizenship.* Multilingual Matters.

Cao, Y., & Philp, J. (2006). Interactional context and willingness to communicate: A comparison of behavior in whole class, group and dyadic interaction. *System, 34*(4), 480–493. https://doi.org/10.1016/j.system.2006.05.002

Chau, T. T. H., & Truong, V. (2018). Integrating cultures into teaching EFL in Vietnam: Teachers' perceptions. *LEARN Journal, 11*(2), 103–105. https://so04.tci-thaijo.org/index.php/LEARN/article/view/161551

Chen, L., Chen, P., & Lin, Z. (2020a). Artificial Intelligence in Education: A review. *IEEE Access, 8*, 75264–75278. https://doi.org/10.1109/ACCESS.2020.2988510

Chen, X., Xie, H., & Hwang, G.-J. (2020b). A multi-perspective study on artificial intelligence in education: Grants, conferences, journals, software tools, institutions, and researchers. *Computers and Education: Artificial Intelligence, 1*, Article 100005. https://doi.org/10.1016/j.caeai.2020.100005

Conneau, A., Lample, G., Ruder, S., & Johnson, M. (2020). Unsupervised cross-lingual representation learning. *Proceedings of the 2020 Conference on Empirical Methods in Natural Language Processing (EMNLP 2020).* https://doi.org/10.48550/arXiv.1911.02116

Dizon, G. (2017). Using intelligent personal assistants for second language learning: A case study of Alexa. *TESOL Journal, 8*(4), 811–830. https://doi.org/10.1002/tesj.353

Dong, Y. (2023). Revolutionizing academic English writing through AI-powered pedagogy: Practical exploration of teaching process and assessment. *Journal of Higher Education Research, 4*(2), 52–52. https://doi.org/10.32629/jher.v4i2.1188

Dörnyei, Z. (2005). *The psychology of the language learner: Individual differences in second language acquisition.* Lawrence Erlbaum Associates.

Fathi, J., Rahimi, M., & Derakhshan, A. (2024). Improving EFL learners' speaking skills and willingness to communicate via artificial intelligence-mediated interactions. *System, 121*, Article 103254. https://doi.org/10.1016/j.system.2024.103254

Gallagher, H. C., & Robins, G. (2015). Network statistical models for language learning contexts: Exponential random graph models and willingness to communicate. *Language Learning, 65*(4), 929–962. https://doi.org/10.1111/lang.12130

Gao, F. (2006). Language is culture–on intercultural communication. *Journal of Language and Linguistics, 5*(1), 58–67.

Gass, S. M. (2003). Input and interaction. In C. Doughty & M. Long (Eds.), *The handbook of second language acquisition* (pp. 224–255). Blackwell. https://doi.org/10.1002/9780470756492.ch9

Ghafar, Z. N., Salh, H. F., Abdulrahim, M. A., Farxha, S. S., Arf, S. F., & Rahim, R. I. (2023). The role of artificial intelligence technology on English language learning: A literature review. *Canadian Journal of Language and Literature Studies, 3*(2), 17–31. https://doi.org/10.53103/cjlls.v3i2.87

Gong, X., Tang, Y., Liu, X., Jing, S., Cui, W., Liang, J., & Wang, F. Y. (2020). K-9 artificial intelligence education in Qingdao: Issues, challenges and suggestions. In *IEEE international conference on networking, sensing and control (ICNSC)* (pp. 1–6). IEEE. https://doi.org/10.1109/ICNSC48988.2020.9238087

Graesser, A. C. (2016). Conversations with AutoTutor help students learn. *International Journal of Artificial Intelligence in Education, 26*(1), 124–132. https://doi.org/10.1007/s40593-015-0086-4

Guilherme, A. (2019). AI and education: The importance of teacher and student relations. *AI & Soietyc, 34*, 47–54. https://doi.org/10.1007/s00146-017-0693-8

Hamadouch, A., & Aoumeur, H. (2024). Teachers' perceptions towards assessing learners' intercultural communicative competence in the EFL classroom in Algeria. *Jurnal Arbitrer, 10*(4), 384–390. https://doi.org/10.25077/ar.10.4.384-390.2023

Hejazi, S. Y., Sadoughi, M., & Peng, J. (2023). The structural relationship between teacher support and willingness to communicate: The mediation of L2 anxiety and the moderation of growth language mindset. *Journal of Psycholinguistic Research, 52*, 2955–2978. https://doi.org/10.1007/s10936-023-10026-9

Henry, A., & MacIntyre, P. D. (2024). *Willingness to communicate, multilingualism and interactions in community contexts.* Multilingual Matters.

Hoa, C. T. H. (2020). Integrating culture into teaching EFL in general education: A context of Vietnam. *Language Related Research, 11*(5), 227–252. https://doi.org/10.29252/LRR.11.5.227

Hoang, V. Q. & Bui, H. P. (2023). Encouraging EFL students' willingness to communicate inside Vietnamese high school classrooms: Teachers' strategies and students' beliefs. *Applied Research on English Language, 12*(2), 19–44. https://doi.org/10.22108/ARE.2022.134674.1968

Huang, L., Li, M., & Lee, C. (2020). Neural machine translation for low-resource languages: A review. *Journal of Artificial Intelligence Research, 69*, 593–617. https://doi.org/10.1613/jair.1.11873

Ismailov, M. (2021). Virtual exchanges in an inquiry-based learning environment: Effects on intra-cultural awareness and intercultural communicative competence. *Cogent Education, 8*(1), 1982601. https://doi.org/10.1080/2331186X.2021.1982601

Jeon, J., Lee, S., & Choe, H. (2023). Beyond ChatGPT: A conceptual framework and systematic review of speech-recognition chatbots for language learning. *Computers & Education, 206*, Article 104898. https://doi.org/10.1016/j.compedu.2023.104898

Juhra, M.-D., & Palma, R. C. (2024). Speaking anxiety and intercultural competence as predictors of learner's willingness to communicate. *Asian Journal of Language, Literature and Culture Studies, 7*(3), 464–473. https://doi.org/10.9734/ajl2c/2024/v7i3199

Kandlhofer, M., Steinbauer, G., Hirschmugl-Gaisch, S., & Huber, P. (2016). Artificial intelligence and computer science in education: From kindergarten to university. In *IEEE Frontiers in education conference (FIE)* (pp. 1–9). IEEE. https://doi.org/10.1109/FIE.2016.7757570

Khajavy, G. H., Ghonsooly, B., Fatemi, A. H., & Choi, C. W. (2016). Willingness to communicate in English: A microsystem model in the Iranian EFL classroom context. *TESOL Quarterly, 50*(1), 154–180. https://doi.org/10.1002/tesq.204

Khasawneh, M. A. S. (2023). The potential of AI in facilitating cross-cultural communication through translation. (2023). *Journal of Namibian Studies: History Politics Culture, 37*, 107–130. https://doi.org/10.59670/jns.v37i.4654

Klimova, B., & Chen, J. H. (2024). The impact of AI on enhancing students' intercultural communication competence at the university level: A review study. *Language Teaching Research Quarterly, 43*, 102–120. https://doi.org/10.32038/ltrq.2024.43.06

Kohnke, L., Moorhouse, B. L., & Zou, D. (2023). ChatGPT for language teaching and learning. *RELC Journal, 54*(2), 537–550. https://doi.org/10.1177/00336882231162868

Krashen, S. (1982). *Principles and practice in second language acquisition.* Prentice-Hall International.

Le, P. H. H. (2006). Mediation through the first language: A sociocultural study of group work in Vietnam. *RELC Journal, 37*(1), 105–121. https://doi.org/10.1177/0033688206063477

Le, V. C. (2018). A critical analysis of moral values in Vietnam-produced EFL textbooks for upper secondary schools. In H. P. Widodo, M. R. Perfecto, L. V. Canh, & A. Buripakdi (Eds.), *Situating moral and cultural values in ELT materials—The Southeast Asian context* (pp. 111–129). Springer.

Lee, J. S., & Hsieh, J. C. (2019). Affective variables and willingness to communicate of EFL learners in in-class, out-of-class, and digital contexts. *System, 82*, 63–73. https://doi.org/10.1016/j.system.2019.03.002

Li, X., & Zhang, Y. (2022). Cross-lingual language models for low-resource language translation. *Computational Linguistics, 48*(2), 321–345. https://doi.org/10.1162/coli_a_00435

Liddicoat, A. J., & Scarino, A. (2013). *Intercultural Language Teaching and Learning* (2 ed.). Wiley-Blackwell. https://doi.org/10.1002/9781118482070

Liu, G., & Ma, C. (2023). Measuring EFL learners' use of ChatGPT in informal digital learning of English based on the technology acceptance model. *Innovation in Language Learning and Teaching, 18*, 125–138. https://doi.org/10.1080/17501229.2023.2240316

Liu, C., Wei, Z., & Wang, Y. (2021). Advancements in neural machine translation: Techniques and challenges. *IEEE Transactions on Neural Networks and Learning Systems, 34*(4), 567–582. https://doi.org/10.1109/TNNLS.2021.3060678

Long, M. H. (1983). Native speaker/non-native speaker conversation and the negotiation of comprehensible input. *Applied Linguistics, 4*(2), 126–141. https://doi.org/10.1093/applin/4.2.126

MacIntyre, P. D., & Legatto, J. J. (2011). A dynamic system approach to willingness to communicate: Developing an idiodynamic method to capture rapidly changing affect. *Applied Linguistics, 32*(2), 149–171. https://doi.org/10.1093/applin/amq037

MacIntyre, P. D., Dörnyei, Z., & Clément, R. (1998). Conceptualizing willingness to communicative in a L2: A situational model of L2 confidence and affiliation. *The Modern Language Journal, 82*(4), 545–562. https://doi.org/10.1111/j.1540-4781.1998.tb05543.x

McCallum, L. (2024). New takes on developing intercultural communicative competence: Using AI tools in telecollaboration task design and task completion. *Journal for Multicultural Education, 18*(1–2), 153–172. https://doi.org/10.1108/JME-06-2023-0043

Menezes, E., & Juan-Garau, M. (2015). English learners' willingness to communicate and achievement in CLIL and formal instruction contexts. In M. Juan-Garau & J. Salazar-Noguera (Eds.),

*Content-based language learning in multilingual educational environments* (pp. 221–236). Springer.

Miauw, C., & Guo, Y. (2021). Taiwanese EFL learners English proficiency, intercultural competence, and willingness to communicate. *Journal on English as a Foreign Language, 11*(2), 227–249. https://doi.org/10.23971/jefl.v11i2.3126

Mohamed, Y. A., Kannan, A., Bashir, M., Mohamed, A. H., Adiel, M. A., & Elsadig, M. A. (2024). The impact of artificial intelligence on language translation: A review. *IEEE Access, 12*, 25553–25579. https://doi.org/10.1109/ACCESS.2024.3366802

Munandar, M. I. (2019). *The beliefs and practices of Indonesian high-school EFL teachers regarding the teaching and learning of culture and interculturality* [Doctoral dissertation, Victoria University of Wellington]. https://doi.org/10.26686/wgtn.17139614

Mystkowska-Wiertelak, A., & Pawlak, M. (2017). *Willingness to communicate in instructed second language acquisition: Combining a macro- and micro perspective.* Multilingual Matters.

Ng, D. T. K., Leung, J. K. L., Chu, S. K. W., & Qiao, M. S. (2021). Conceptualizing AI literacy: An exploratory review. *Computers and Education: Artificial Intelligence, 2*, Article 100041. https://doi.org/10.1016/j.caeai.2021.100041

Nguyen, M. H. (2013a). EFL students' reflections on peer scaffolding in making a collaborative oral presentation. *English Language Teaching, 6*(4), 64–73. https://doi.org/10.5539/elt.v6n4p64

Nguyen, T. L. (2013). *Integrating culture into Vietnamese University EFL Teaching: A critical ethnography study* [PhD. Thesis, Auckland University of Technology]. New Zealand.

Novita, D., & Purwati, O. (2021). Incorporating culture and textbook adaptation practice for promoting intercultural competence in ELT: Teachers' perspectives. *JEES (Journal of English Educators Society), 6*(1), 96–104. https://doi.org/10.21070/jees.v6i1.1059

Ou, A. W., & Malmström, H. (2023). It becomes increasingly complex to deal with multiple channels: Materialised communicative competence and digital inequality in English-medium higher education in the digital era. *Journal of Multilingual and Multicultural Development.* https://doi.org/10.1080/01434632.2023.2222102

Peng, J. (2012). Towards an ecological understanding of willingness to communicate in EFL classrooms in China. *System, 40*(2), 203–213. https://doi.org/10.1016/j.system.2012.02.002

Peng, J., & Woodrow, L. J. (2010). Willingness to communicate in English: A model in Chinese EFL classroom context. *Language Learning, 60*(4), 834–876. https://doi.org/10.1111/j.1467-9922.2010.00576.x

Peng, J. E., & Liang, W. (2025). Willingness to communicate with artificial intelligence (AI)? Insights from tracking EFL learners' perceived acceptance and chat output. *Computer Assisted Language Learning.* (Advance online publication). https://doi.org/10.1080/09588221.2025.2486147

Peng, J., & Wu, L. (2022). Motivational profiles of Chinese university students majoring in Spanish: A comparative study. *Journal of Multilingual and Multicultural Development,* Advance online publication. https://doi.org/10.1080/01434632.2022.2035740

Peng, J. (2014). *Willingness to communicate in the Chinese EFL university classroom: An ecological perspective.* Multilingual Matters.

Philp, J., Adams, R., & Iwashita, N. (2013). *Peer interaction and second language learning (1st ed.).* Routledge. https://doi.org/10.4324/9780203551349

Pires, T., Schlinger, E., & Garrette, D. (2019). How multilingual is multilingual BERT? *Proceedings of the 2019 Conference on Empirical Methods in Natural Language Processing (EMNLP 2019).* https://doi.org/10.48550/arXiv.1906.01502

Rahimi, M., & Fathi, J. (2024). Employing e-tandem language learning method to enhance speaking skills and willingness to communicate: The case of EFL learners. *Computer Assisted Language Learning, 37*(4), 924–960. https://doi.org/10.1080/09588221.2022.2064512

Robinson, S. C. (2020). Trust, transparency, and openness: How inclusion of cultural values shapes Nordic national public policy strategies for artificial intelligence (AI). *Technology in Society, 63*, Article 101421. https://doi.org/10.1016/j.techsoc.2020.101421

Rodríguez-García, J. D., Moreno-Leon, J., Román-González, M., & Robles, G. (2020). Introducing artificial intelligence fundamentals with LearningML: Artificial intelligence made easy. *Eighth International Conference on Technological Ecosystems for Enhancing Multiculturality* (pp. 18–20). ACM. https://doi.org/10.1145/3434780.3436705

Saito, K., & Hanzawa, K. (2015). Developing second language oral ability in foreign language classrooms: The role of the length and focus of instruction and individual differences. *Applied Psycholinguistics, 37*(4), 813–840. https://doi.org/10.1017/S0142716415000259

Sarwari, A. Q., Javed, M. N., Adnan, H. M., & Wahab, M. N. A. (2024). Assessment of the impacts of artificial intelligence (AI) on intercultural communication among postgraduate students in a multicultural university environment. *Scientific Reports, 14*, 13849. https://doi.org/10.1038/s41598-024-63276-5

Sato, C. J. (1990). Ethnic styles in classroom discourse. In M. Hines & W. Rutherford (Eds.), *Or TESOL 81* (pp. 11-24). Teachers of English to Speakers of Other Languages.

Tai, T. Y., & Chen, H. H. J. (2020). The impact of Google Assistant on adolescent EFL learners' willingness to communicate. *Interactive Learning Environments, 31*(3), 1485–1502. https://doi.org/10.1080/10494820.2020.1841801

Tai, T. Y. (2022). Effects of intelligent personal assistants on EFL learners' oral proficiency outside the classroom. *Computer Assisted Language Learning*, 1–30. https://doi.org/10.1080/09588221.2022.2075013

Takahashi, A. (2009). Self-perception of English ability: Is it related to proficiency and/or class performance? Doctoral dissertation, Niigata University, Japan.

Vaswani, A., Shazeer, N., Parmar, N., Uszkoreit, J., Jones, L., Gomez, A. O., Kaiser, Ł., & Polosukhin, I. (2017). Attention is all you need. *Proceedings of the 31st Conference on Neural Information Processing Systems (NeurIPS 2017)*. https://doi.org/10.48550/arXiv.1706.03762

Vygotsky, L. S. (1978). *Mind in society: The development of higher psychological processes.* Harvard University Press.

Wei, W., & Cao, Y. (2021). *Willing, silent or forced participation?* RELC Journal, Advance online publication. https://doi.org/10.1177/00336882211066619

Wong, G. K., Ma, X., Dillenbourg, P., & Huan, J. (2020). Broadening artificial intelligence education in K-12: Where to start? *ACM Inroads, 11*(1), 20–29. https://doi.org/10.1145/3381884

Yates, L., & Trang, N. T. Q. (2012). Beyond a discourse of deficit: The meaning of silence in the international classroom. *International Education Journal: Comparative Perspectives, 11*(1), 22–34.

Zhang, T., Lu, E., Liao, Q., & Sun, D. (2025). A study of college students' academic anxiety modeling based on interpretable machine learning algorithms. *Journal of Psychoeducational Assessment, 43*(4), 437–450. https://doi.org/10.1177/07342829251317251

Zhang, C., Meng, Y., & Ma, X. (2024). Artificial intelligence in EFL speaking: Impact on enjoyment, anxiety, and willingness to communicate. *System, 121*, Article 103259. https://doi.org/10.1016/j.system.2024.103259

Zou, B., Wang, C., Yan, Y., Du, X., & Ji, Y. (2024). Exploring English as a foreign language learners' adoption and utilisation of ChatGPT for speaking practice through an extended technology acceptance model. *International Journal of Applied Linguistics*. https://doi.org/10.1111/ijal.12658

Zou, B., Xie, S., & Wang, C. (2025). Students' willingness to communicate (WTC) in using artificial intelligence (AI) technology in English-speaking practice. *International Journal of Information and Communication Technology Education, 21*(1), 1–18. https://doi.org/10.4018/IJICTE.375387

**Thanh Tien Nguyen** is currently a graduate student in English Language Studies at Hoa Sen University, Vietnam. His interests mainly focus on academic English and test preparation, including IELTS and TOEIC, diagnosing students' needs and learning difficulties. He also conducts training sessions for L2 teachers on assessment practices and evaluates their assessment competence.

**Hung Phu Bui** is an associate professor at Faculty of Foreign Languages, Ton Duc Thang University, Vietnam. He holds a PhD in English Language Education. His research interests span across many aspects of applied linguistics and language education: second language acquisition, L2 classroom assessment, teaching English for specific purposes, and computer-assisted language teaching and learning. Influenced by educational, linguistic, and psychological perspectives, his endeavors, mainly published in leading journals in the fields of language education and applied linguistics, have been stimulating discussions. Serving as the keynote and plenary speaker in many national and international conferences, Hung has had opportunities to spread his knowledge and research interests to students, colleagues, and novice researchers.

# Artificial Intelligence–Enhanced Scaffolding and Peer Collaboration in Culturally and Linguistically Diverse Classrooms: Evidence from Asia

**D. Solomon Paul Raj, K. Saritha, Hariharasudan Anandhan, Mari Carmen Gil Ortega, S. Selvalakshmi, and M. Gopinath**

**Abstract** This experimental study compares the effects of scaffolding via artificial intelligence (AI) and peer scaffolding on learning engagement in culturally and linguistically diverse classrooms in Asia. It used an explanatory sequential mixed-methods design with data from 280 South Asian students (aged 14–18) from 12 schools in India, Pakistan, and Bangladesh. Quantitative data were collected using the Learning Engagement Scale, AI Scaffolding Effectiveness Questionnaire, and Peer Collaboration Quality Index. Qualitative data collection included semi-structured interviews and classroom observations Quantitative data were analyzed using ANOVA, multiple regression, principal component analysis, k-means, and random forest modelling. PCA analysis indicated components of engagement as AI-Scaffolded Engagement (42.3%), Peer Support Quality (21.1%), and Cultural Responsiveness (16.2%). ANOVA results indicate significant differences between scaffolding conditions ($F(2,277) = 24.6$). PCA analysis revealed components of

D. Solomon Paul Raj (✉) · M. Gopinath
Sri Krishna Arts and Science College, Coimbatore, Tamil Nadu, India
e-mail: solomon.david@skasc.ac.in

M. Gopinath
e-mail: gopinathm@skasc.ac.in

K. Saritha · S. Selvalakshmi
Karpagam Academy of Higher Education, Coimbatore, Tamil Nadu, India
e-mail: sarithamahesh@ymail.com

S. Selvalakshmi
e-mail: selva.lakshmi85@gmail.com

H. Anandhan
Department of Language, Culture and Society, College of Engineering and Technology, SRM Institute of Science and Technology, Kattankulathur, Tamil Nadu, India
e-mail: dr.a.hariharasudhan@gmail.com

M. C. G. Ortega
Department of Education, University of Bath, Bath, England
e-mail: mdcgo20@bath.ac.uk

 
H. P. Bui and L. P. H. Huong (eds.), *Artificial Intelligence in Education and Academic Research*, Studies in Computational Intelligence 1269,
https://doi.org/10.1007/978-3-032-18392-7_9

engagement as AI-Scaffolded Engagement (42.3%), Peer Support Quality (21.1%), and Cultural Responsiveness (16.2%). ANOVA results indicated significant differences among scaffolding conditions (F (2,277) = 24.67, $p < 0.001$, $\eta^2 = 0.15$). Based on random forest models, learning engagement was predicted with 89% accuracy using AI feedback frequency and peers' interaction quality as predictors. Qualitative data emphasised the importance of culturally responsive peer collaboration in supporting multilingual learning. The study shows that AI scaffolding and culturally responsive peer support can optimise engagement with students when taken together, consistent within Zone of Proximal Development.

**Keywords** Zone of proximal development · AI scaffolding · Peer collaboration · Multilingual education · Cultural responsiveness · Educational technology

## 1 Introduction

Education in culturally and linguistically diverse (CLD) classrooms presents both challenges and opportunities, particularly across South Asia, where secondary students often learn in multilingual settings and the medium of instruction may differ from learners' home languages. Globally, an estimated share of students is not taught in a language spoken at home, a pattern associated with disengagement and inclusion concerns (Hill, 2020). Vygotsky's Zone of Proximal Development (ZPD) offers a foundational framework for the scaffolded support examined in this discussion, mediated by the educator and between learner capabilities (Vygotsky, 1980). Scaffolding in multilingual contexts is more complex as it is mediated and shaped by students' repertoire of language and cultural backgrounds and the affordances of technology to support these activities (Hickey & Riddle, 2021).

The advancements in artificial intelligence (AI) offers another layer of dynamic, context-sensitive scaffolding that not only responds to learner needs but preserves peer collaboration as a feature of learning (Hung & Nguyen, 2022). Early evidence shows that, in general, AI tutors and chatbots can support the engagement of behavioral, cognitive and emotional aspects, while collaborative learning approaches (e.g. peer tutoring, cooperative problem-solving, and group discussions) are established methods that enhance motivation, agency and achievement (Omar et al., 2025). In the context of CLD cohorts, the dynamics among peers provide opportunities to be supported and further foster a hospitable, inclusive learning context that supports a sense of learners.

Against this backdrop, the present study examines how AI-enhanced instructional scaffolding and peer collaborative learning jointly support student engagement across multicultural classrooms in India, Pakistan, and Bangladesh (De Brún et al., 2021). Guided by sociocultural theory, we investigate the roles of peer interaction and AI tools as instructional scaffolds that help maintain learners within their ZPD. Although interest in AI for education is rapidly expanding, empirical work on the interaction between AI scaffolding and peer collaboration in South Asian multilingual settings remains limited; prior studies often focus on Western contexts or treat AI and collaboration separately (Bassi et al., 2025). Addressing this gap, we pose four questions:

(1) What impacts does AI-supported scaffolding have on engagement in multilingual milieus?
(2) How does peer collaboration mediate the AI–engagement relationship?
(3) To what extent do cultural and linguistic factors facilitate the efficacy of AI-scaffolded environments?
(4) Which input factors best predict engagement profiles in technology-enhanced multilingual classrooms?

## 2 Literature Review

Vygotsky's ZPD remains a central framework for understanding learning in culturally diverse contexts such as India and South Asia. Extensions of ZPD such as the Zone of Collaborative Development (ZCD) highlight the dialogic, peer-driven dimensions of learning evident in multicultural classrooms (Hockham & Wallis, 2023). Evidence from South Asian mathematics education shows that misconceptions can persist when instruction is not scaffolded within students' proximal capabilities, especially among linguistically and socially diverse learners (Cahyono & Mukin, 2025). Cultural-historical activity theory also argues that ZPD should be contextualized within learners' particular social contexts and "leading activities" consistent with the pluralist philosophy of Indian classrooms (Sagredo-Gallardo et al., 2025). In higher education contexts, ZPD-informed approaches can provide insights into how disadvantaged or first-generation learners negotiate new academic spaces (Zhang & Yin, 2025). Additionally, the development of theoretical concepts such as zones of prompted action and variable development have exemplified the importance of adaptive, instructional responses to the realities of being a student in a South Asian classroom (Shahini, 2025).

### *2.1 Theoretical Foundations of the ZPD*

The term ZPD traditionally refers to the difference between an independent scholar and knowledgeable peers (Yan, 2025). Hence, learning is socially mediated. Scaffolding, therefore, is a temporary, structured support moving towards independence. Research evidence has presented different scaffolds in action cognitively, metacognitively and linguistically but perhaps most important in a multilingual classroom, is the idea of using dialogic interaction, peer modelling and responsive feedback (Lantolf et al., 2018). The arguments presented here help establish a coherent framework for adapting pedagogy for CLD learners, with a cognitive, linguistic and cultural framework interspersed (Bruner, 1984).

## 2.2 *Cultural Dimensions of Learning and ZPD*

AI-based instructional tools, such as adaptive tutors, chatbots, and analytics-informed feedback systems, can serve as interactive artifacts scaffolding learners to remain in the appropriate range in the Zone of Proximal Development (ZPD) (Hickey & Riddle, 2021). This suggests that scaffolding might involve AI detection of patterns in learner responses and timing, so that cognitive prompts, metacognitive prompts, and linguistic supports (e.g., glosses, paraphrases, code-switching suggestions) can be adjusted in real time, which is especially important in culturally and linguistically diverse (CLD) environments (Yan et al., 2025). Evidence from early pilot studies suggests that when learners receive contingent support from a dialogic agent, their engagement and comprehension improve, with the principles of ZPD and learner agency (Puntambekar, 2021).

In multilingual classrooms, AI-based scaffolds might augment teacher-mediated instructional practices by: (i) providing just-in-time supports regarding language (such as morphological cues, vocabulary retrieval, bilingual genre-based frames, etc.); (ii) pacing cognitive load through task-step decomposition; and, (iii) stimulating reflexive think-aloud via metacognitive prompts (Hung, 2019a, 2019b; Martin-Beltrán et al., 2017). These scaffold supports will likely be most effective when embedded in tasks that focus on collaboration as peers interact, with machine guidance that enhances rather than displaces collaborative work (Hung, 2019a, 2019b; Scotland, 2022). From a design perspective, there is a need for guardrails to avoid "over scaffolding," where hints defeat productive struggle, but the scaffolding elicited by the intervention is adaptive and teacher humane overrides or ways to reduce hints support broader instructional principle of ZPD learning (Lee et al., 2024).

## 2.3 *Peer Collaboration and Scaffolding in Diversely Linguistic Classrooms*

The research connects collaborative structures to enhancements in achievement, motivation, and social integration outcomes, which is especially relevant in heterogeneous classes (Sampson, 2021). In CLD students' classrooms, peer dialogue can facilitate access to both disciplinary language and practices, including situated explanations and examples that connect with students' cultural repertoires. When collaboration is intentionally designed (e.g. clear roles, interdependence, accountability), it supports equitable participation and combats marginalization, enhancing both belonging and persistence in the classroom (Storch, 2005).

In the case of multilingual students' classrooms, collaboration supports linguistic scaffolding: peers model terminology, syntax, and discourse move when co-constructing meaning around tasks; all this work occurs in relation to ZPD perspectives of social mediation and the ethics of learning (Hu et al., 2024). Engaging in collaboration, however, should not be viewed as the elixir; without an intentional structure, status differences can become embedded under the guise of collaboration, or responsibility can be spread too thin. Evidence suggests that maximizing gains in engagement and conceptual understanding requires some form of cooperative design (e.g. positive interdependence, rotating roles) along with either a teacher or AI prompt (Hamadouch & Aoumeur, 2024; Smith et al., 2024).

## *2.4 Technology-Mediated ZPD in Asia*

Prior work establishes that (a) scaffolding calibrated to the ZPD promotes learning and engagement, and (b) well-designed peer collaboration enhances motivation, participation, and achievement in diverse classrooms (Van Le et al., 2025). The studies commonly examine AI scaffolding and collaboration separately, or in Western contexts, with limited attention to how their interaction operates in South Asian multilingual settings (Reitman et al., 2023). This leaves an important gap: whether AI-mediated guidance can augment collaborative processes (rather than compete with them) to sustain engagement when language, culture, and prior schooling vary widely. Building on this synthesis, we articulate four testable hypotheses for CLD secondary classrooms:

H1 (AI → Engagement): AI-enhanced scaffolding will be positively associated with behavioral, cognitive, and emotional engagement, controlling baseline characteristics.

H2 (Peer Mediation): The relationship between AI scaffolding and engagement will be partially mediated by the quality of peer collaboration (e.g., equitable turn-taking, elaborative talk) (Isaac et al., 2023).

H3 (Context Moderation): Cultural-linguistic factors (e.g., home–school language match, prior exposure to English/LoI) will moderate the effect of AI scaffolding on engagement, with stronger effects where linguistic distance is greater and supports are explicit (Amaya et al., 2023).

These hypotheses posit that AI is not as a substitute for human interaction but as a complementary scaffolding tool that can stabilize collaborative learning within the ZPD in CLD environments, addressing the documented gap in multilingual, Global South contexts (Henss & Pinquart, 2023).

# 3 Research Methodology

## 3.1 Research Design

The aim of this research was to examine the impact of AI-supported scaffolding and peer collaboration on learner engagement in multilingual classrooms (García-Holgado et al., 2024). In two phases the study was conducted. The quantitative phase, was investigated first, it elucidates the relationship between key variables and hypotheses with the ZPD theory. The second, qualitative phase, was an explanation of the previous relationships that had been examined, placing emphasis on the processes and cultural contexts in which they took place. The explanatory sequential mixed methods design was justified because it drew on the strengths of both quantitative and qualitative approaches: using quantitative research to identify relationships and patterns and then explaining these through qualitative research in a process context that included cultural context that was influencing the learning experience (Yan, 2025).

## 3.2 Participants and Setting

### 3.2.1 Sample Size and Power Analysis

Using G*Power 3.1.9.7 software, a power analysis showed that 264 student participants were necessary to ensure obtaining 80% statistical power when detecting medium effect sizes ($f = .25$) for ANOVA analyses (Wen et al., 2024) ($\alpha = 0.05$). The researcher aimed to recruit 280 participants to account for potential attrition and ensure enough statistical power for a follow-up machine learning analysis.

### 3.2.2 Participant Characteristics

The sample for this research included 280 students aged 14 to 18 years ($M = 16.2$; $SD = 1.3$) from 12 secondary schools in India ($n = 168$), Pakistan ($n = 72$), and Bangladesh ($n = 40$). Among these students, 142 were female (50.7%) and 138 were male (49.3%). The students from the schools were selected through a stratified random sampling procedure, including urban (58%), semi-urban (26%), and rural (16%) representation (Wang et al., 2022). We employed the following criteria to select the participants.

- The participants were in high school grades (e.g., 9–12)
- They possess basic digital literacy, as indicated by their report of being comfortable with using the computer or tablet

- They were competent in more than one language, (i.e., be fluent in at least two languages)
- They demonstrate both participant and guardian informed consent to participation.
- They were excluded from the data collection if they met any specified criteria:
- They did not have any learning disabilities requiring specialised intervention
- They fully attended the intervention period.

The missing data in more than 50% of the study measures.

## 3.3 *Intervention Design*

### 3.3.1 AI-Enhanced Scaffolding Platform

The action taken used Edu Scaffold 2.0, an adaptive AI tutoring platform designed for multilingual education. The platform provided three general types of scaffolding: (1) cognitive scaffolding via adaptive hint sequences, (2) metacognitive scaffolding through self-regulation prompts, and (3) linguistic scaffolding, including multilingual assistance and translation support (Wu & Peng, 2025).

### 3.3.2 Peer Collaboration Structure

Students were placed in heterogeneous groups of three to four members, with attention paid to linguistic diversity and prior academic performance in group composition. Collaborative approaches included joint problem solving, peer tutoring, and tasks soliciting multilingual varieties of explanation. To diversify the exposure to collaboration styles and linguistic practices, group membership was rotated weekly, in alignment with the process (Pagani, 2021).

### 3.3.3 Experimental Conditions

The three experimental conditions experienced by the participants were: (1) AI-scaffolding alone ($n = 94$), (2) peer collaboration only ($n = 93$), or (3) AI-scaffolding and peer collaboration ($n = 93$). Each of the three groups received curriculum content delivered digitally for eight weeks (Tarifa-Rodriguez et al., 2024).

## 3.4 Data Collection Instruments

### 3.4.1 Quantitative Measures

Learning Engagement Scale (LES): This 24-item tool assessed engagement in three domains cognitive (8), emotional (8), and behavioural (8)—using a 5-point Likert Scale The scale demonstrated good reliability ($\alpha = 0.89$) and was developed for use in multiple languages (Ornelles, 2007). Example items are: "I focus my full attention to completing the tasks to learn." (cognitive domain) and "I feel enthusiastic to learn new things." (emotional domain).

Those samples included a questionnaire of AI Scaffolding Effectiveness Questionnaire (ASEQ) with 20 items that evaluated the perceived utility (10 items) and ease of interaction (10 items) of AI scaffolding attributes, demonstrating excellent reliability ($\alpha = 0.92$) and factorial validity as evidenced by confirmatory factor analysis (CFI = 0.96, RMSEA = 0.04). Example items included: "The AI Systems provided hints when I was stuck." and "The AI feedback was easy to understand."

Peer Collaboration Quality Index (PCQI) was an eighteen-item scale that evaluated effectiveness of collaboration across communication (6 items), coordination (6 items), and mutual support (6 items). The tool also showed evidence of good reliability ($\alpha = 0.87$) and factorial structure ($\chi^2 = 124.33$, $df = 132$, $p = 0.67$). Example items are: "My group members explain concepts clearly to one another," and "We work well together to solve problems." The assessment of academic performance consisted of standardised tests to source conceptual understanding and problem-solving ability that aligned to the objectives and content of the curriculum covered during the intervention.

### 3.4.2 Qualitative Data

Twenty four of the participants were raondomly selected for interviews of 30 to 45 min each (Tin & Bui, 2024). The interviews aimed to explore their experiences with AI scaffolding, peer collaboration, and other elements of culture in relation to learning. The interview protocol included specific questions about the participants' learning preferences, experiences with technology (and culture), and other aspects of group work. Semi-structured interviews were used to allow for the opportunity to gather an in-depth understanding of the participant's estimates, while allowing for flexibility to question other matters about emerging answers in participants' answers to the interview protocol questions (Bui & Vo, 2024). Semi-structured interviews were used to gain the nuances needed about issues that were differently with technology and culture that shape some aspects of learning (Ornelles, 2007).

Classroom Observations: Structured observations utilising the Classroom Assessment Scoring System (CLASS) adapted for technology-enhanced learning situations were used to capture students' behaviours, interactional patterns, and culturally situated practices engaged with a group task, as a whole, there were 96 h observing in total per condition and school. The CLASS captures timely rich, contextual data about group dynamics and culturally situated practices. CLASS was used with an understanding of gathering authentic data based on classroom experiences that could not have been obtained through interviews (Burch et al., 2015).

## 3.5 *Data Collection*

### 3.5.1 Timeline and Protocol

Data collection took place over a 12-week period and followed a prescribed protocol. During week 1, we conducted the baseline assessments, and students were orientated to the platform. Weeks 2–9 constituted the intervention phase, within which students were measured week on week. Weeks 10–12 involved the post-intervention assessments, student interviews and final observations.

### 3.5.2 Quality Assurance

Research assistants were trained for 16 h on the data collection protocol, cultural sensitivity, and technology troubleshooting. Inter-rater reliability on observational coding was > 85% agreement ($\kappa = 0.82$). The platform usage data logged electronically to verify fidelity of the intervention. These measures served to establish data reliability and validity necessary to support inferences about the effectiveness of the intervention.

## 3.6 *Data Analysis Plan*

### 3.6.1 Quantitative Analysis

Descriptive statistics were used to characterise sample demographics and variable distributions, providing essential insight into the study population. Assumption testing included normality (Shapiro–Wilk tests), homogeneity of variance (Levene's tests), and multicollinearity testing (VIF < 5.0), all of which demonstrate that the chosen analytical methods are appropriate for interpretation of the data.

ANOVA was chosen to test group differences in engagement across experimental conditions because it provides a means of statistically comparing means between two or more groups, provides a measure of robust analysis of multiple groups, and directly addresses the research question about the comparative effects of AI scaffolding, PEER collaboration and both AI scaffolding & PEER collaboration on engagement.

Multiple regression was used to test predictors of learning engagement, and to examine significant factors of engagement and relevant contributions of each factor, which aligns with the goal of revealing drivers of engagement. Principal Component Analysis (PCA) was employed to identify positions of engagement (i.e., dimensions of engagement) that align with the study's conceptual framework, reducing a multiplicity of engagement data to components that can be interpreted. K-means clustering was employed to classify student engagement profiles, capturing the possibility of natural subgroups that react differently. Random Forest modelling was employed to test whether engagement would be predicted and to identify variable importance, providing a powerful non-linear analysis for identifying key predictors in complex education data (Kuiper et al., 2025).

Effect sizes were calculated using Cohen's conventions (small = 0.2, medium = 0.5, large = 0.8 for d; small = 0.01, medium = 0.06, large = 0.14 for $\eta^2$) to support reasonable interpretation of statistical analysis. Missing data (<5% across all measures), were retained by using multiple imputation for mean values across the data set using the MICE algorithm, allowing for power of inferential analysis with less risk of bias.

### 3.6.2 Qualitative Analysis

Reflexive thematic analysis was employed to analyse the transcripts from the interviews. Reflexive thematic analysis was selected because it effectively conveys the depth and richness of participants' experiences, which is an essential requirement for understanding the role of complex social and cultural influences on learning. Two researchers independently coded the transcripts, and inter-coding reliability was 78% at initial coding, and this improved to 92% after conversation and a process of independent and jointly discussing and refining codes. Codes were categorised into themes using a constant comparative analysis, which aligned with the study's aim of portraying a complex account of the data. NVivo 12 was used to support data management and coding (Liu & Luo, 2025).

### 3.6.3 Integration Strategy

This was done in two ways: comparing the findings from statistical results with the qualitative thematic findings, along with creating some meta-inferences looking for convergence and divergence between the various data sources (Fetters & Guetterman, 2021). Joint displays will visualise the quantitative engagement scores alongside qualitative themes and interview quotes extracted from the data analysis, facilitating direct comparison and exploring convergent or divergent patterns of behaviours. Meta-inferences will then be developed through ongoing team discussions to synthesise thinking across the data sources, enhancing the validity and explanatory power of the mixed-methods components of this study.

### 3.7 Ethical Considerations

Joint displays were used to bring quantitative and qualitative findings together by comparing statistical results with thematic findings, as well as through meta-inferences that investigated convergence and divergence across the various data sources (Su et al., 2022). Joint displays will visualise quantitative engagement scores alongside qualitative themes and interview quotes extracted from the data analysis, facilitating direct comparison and exploring convergent or divergent patterns of behaviour. Meta-inferences will then be developed through ongoing team discussions to synthesise thinking across the data sources enhancing the validity and explanatory power of the mixed-methods components of this study.

## 4 Results and Discussion

### 4.1 Overview

The outcomes are conveyed in two main sections: (1) quantitative analyses to investigate associations across AI enabling scaffolding, peer collaboration and learning outcomes; and (2) qualitative analyses to identify emergent patterns in participants' experiences. Taken together, these results show the interactions among cultural, linguistic, and technological aspects of the Zone of Proximal Development.

### 4.2 Descriptive Statistics

Descriptive statistics for the main study variables, AI-Enhanced Scaffolding, Peer Collaboration, and Learning Outcomes, are presented in Table 1. In regard to scaffolding, the mean score was M = 4.18; SD = 0.62 indicating a fairly high perceived usefulness of AI-mediated scaffolding. The means for peer collaborating suggested a one-time M = 4.02; SD = 0.68, and for learning outcome, it was M = 78.50; SD = 12.7 out of (100).

**Table 1** Descriptive statistics (**n = 280**) with Cronbach's α

| Variable | Mean (M) | SD | Min | Max | Cronbach's α |
|---|---|---|---|---|---|
| AI-enhanced scaffolding | 4.18 | 0.62 | 2.6 | 5 | 0.87 |
| Peer collaboration | 4.02 | 0.68 | 2.4 | 5 | 0.85 |
| Learning outcomes | 78.5 | 12.7 | 45 | 100 | 0.82 |

Overall, high internal consistency coefficients support the reliability of the measurement instruments, consistent with previous studies on educational scales and scaffolding constructs (Daga & Rajan, 2023).

## 4.3 *Inferential Analysis*

### 4.3.1 Group Differences

The independent samples t-tests showed no significant difference ($p > .05$, Cohen's $d = .12$) in scaffolding, or collaborative scores between male or female students. The results also showed no meaningful difference ($p > .05$) between government and private schools. However, a one-way ANOVA, did show significant differences across language of instruction groups ($F(3,276) = 4.27$, $p = .006$, $\eta^2 = .07$). Students in bilingual or trilingual contexts showed a significantly higher collaboration and experience outcome scores compared to students in monolingual classrooms, as has been established in previous literature that documented the positive benefits of multilingual learning environments (Lin et al., 2024).

### 4.3.2 Correlation Analysis

Pearson correlation coefficients demonstrated positive and significant associations among the three constructs (see Table 2). Strong intercorrelations indicate that scaffolding and collaboration jointly predict improved learning outcomes, consistent with sociocultural learning theories, with $p < 0.01$.

## 4.4 *Predictive Modelling Using AI Algorithms*

To extend the interpretation beyond standard descriptive statistics, 4 machine learning models were used: Principal Component Analysis (PCA), K-Means Clustering, Decision Tree, and Random Forest (Hao et al., 2025).

**Table 2** Pearson correlations among AESS, PCI, LOAT (**n = 280**)

| Variables | 1 | 2 | 3 |
|---|---|---|---|
| 1. AI-enhanced scaffolding | 1 | — | — |
| 2. Peer collaboration | 0.71** | 1 | — |
| 3. Learning outcomes | 0.63** | 0.68** | 1 |

### 4.4.1 PCA and K-Means Clustering

PCA extracted three principal components that accounted for 79.6% of variance explained overall, while component 1 labelled as (AI-Scaffolded Engagement) had significant loadings on the interaction and feedback factors. Component 2 (Peer Support) captured collaborative responses among participants. Component 3 (Outcome Orientation) measured goal-directed learning. Finally, a K-Means clustering procedure (k = 3) was completed to categorise participants into:

Cluster 1: Highly Engaged Learners (42%),

Cluster 2: Moderately Engaged Learners (37%), and

Cluster 3: Low Engagement Group (21%).

The clusters showed considerable overlap between levels of engagement with CAAI and collaboration with peers (see Fig. 1). The PCA results produced three components that explained 79.6% of the total variance and thus demonstrated strong conceptual fit among the variables. Component 1 (AI-Scaffolded Engagement) was associated with a heavy degree of variables associated with adaptive feedback, interactive prompts, and how often learners interacted with the AI, capturing the cognitive and motivational aspects of AI-supported learning. Component 2 (Peer Support) captured collaborative indicators including group learning, joint problem-solving, and dialogue back-and-forth, focusing on the social constructivist dimension of the phenomena of learning. Outcome Orientation (Component 3) was more focused on learning behaviours that were goal oriented, self-monitoring, and task completion, representing the learners' strategic engagement dimension. K-Means classifications (k = 3) were then conducted where learners were placed into three meaningful categories: Cluster 1 (Highly Engaged Learners, 42%) Learners who demonstrated this strong relationship between AI scaffolding and peer collaboration; Cluster 2 (Moderately Engaged Learners, 37%), where there was partial engagement and varied peer interaction; and Cluster 3 (Low Engagement Group, 21%)—minimal AI involvement and engagement with peers. Whereas the clusters differentiated the level of AI-mediated engagement the overlapping nature of the three clusters illustrated the interdependence between AI-mediated engagement and peer collaboration, suggesting that technological scaffolds are more effective supporting learner collaborative meaning making when the context is socially interactive and culturally responsive.

This evidence supports assertion that AI-based learning platforms can visualise learner trajectories in the ZPD to enhance the quality of scaffolding students receive (Wang et al., 2022).

### 4.4.2 Decision Tree and Random Forest Modelling

Decision trees and random forests were used to predict learning outcomes in scaffolding and collaboration (see Fig. 2). The decision tree achieved 82% accuracy, while the random forest achieved 89%, demonstrating similar levels of accuracy. The feature importance analysis showed that frequency of AI feedback as well as peer

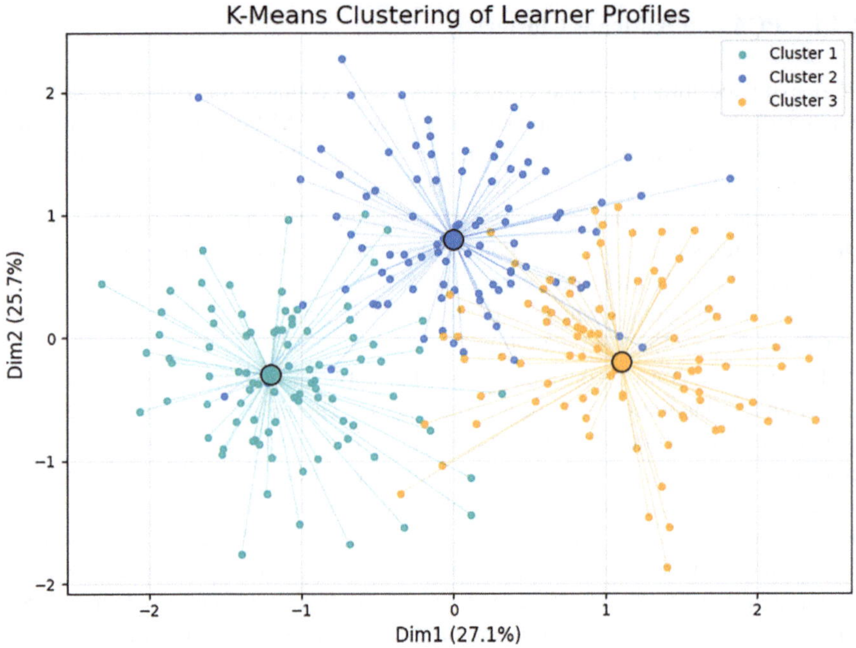


**Fig. 1** PCA biplot on AESS, PCI, LOAT variables with K-Means clusters (k = 3; n = 280)

interaction quality were the strongest predictors. These findings support the value of ensemble learning methods in educational data mining, allowing for the capture of non-linear relationships more effectively than traditional regression models. The predictive capabilities of AI-based models support the assumption that adaptive scaffolding and peer interaction are not individual constructs; rather, they are representative of two dynamic co-determined factors that influenced learners' cognitive and social engagement within a given classroom ecosystem (Hmoud et al., 2025). The findings support the results by Omar et al. (2025), who emphasised that the addition of AI to a learning support tool greatly improved the predictive accuracy of learning in diverse South Asian classroom learning contexts.

### *4.5 Qualitative Findings*

Using thematic analysis to analyse the interview data resulted in three main themes. First, AI Mediation via Adaptive Scaffolding: AI tools provided students with adaptive and a greater variety of feedback/responsiveness. This is closely associated with the scaffolding framework articulated by (Omar et al., 2025). Students reported that the AI mediation-enhanced systems provided personalised scaffolding, adapting the difficulty of the problem to their readiness, and consequently, promoting their

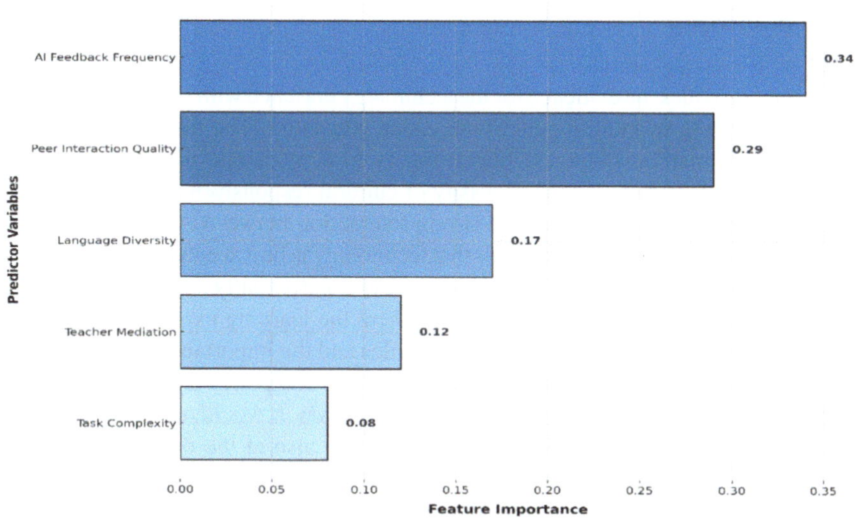


**Fig. 2** Random forest feature importance predicting LOAT from AESS/PCI

problem-solving skill development (Celik et al., 2024). The second theme is about collaborative knowledge construction. The students engaged with expressions from peers, which align with various theorists related of dialogic learning practices. Many of the students said that the AI (through prompts) helped to create more opportunities to develop agency with their multilingual peers (Celik et al., 2024). The final theme is about cultural inclusivity and motivation. Participants noted that the AI reduced language barriers through multilingual options, which is an important aspect for fostering an inclusive learning experience (Zhang & Yin, 2025). This inclusivity extended to supporting diverse learning styles and cultural backgrounds, thereby enhancing overall motivation and engagement within the educational environment.

## *4.6 Integration of Quantitative and Qualitative Results*

The synthesis of the two data strands revealed that AI-enhanced scaffolding plays a positive mediating role in peer collaboration, which predicts learning performance. Thus, confirming the theoretical model of reciprocal interaction within the ZPD. The triangulated nature of the findings also validates the assertion that AI-based environments serve as mediating artefacts that promote communication, coordination, and the co-construction of knowledge.

## 5 Conclusion

This study provides the evidence for the technology enriched with peer collaboration-aided scaffolding in multilingual South Asian education. The findings expand the Zone of Proximal Development (ZPD) theory to demonstrate not only the utility of such technology but also the necessity of cultural responsiveness to offer educational technology. Furthermore, the strong interaction between AI scaffolding and learning from peers also demonstrates that technology is best at enhancing rather than replacing human relations and interactions. This is particularly significant in culturally and linguistically diverse classrooms where the learning experience is created by human products. Identifying learning profiles and the importance of multilingual support, give direction to current and future teachers and policy makers who continually learn and advance through digital transformation. It would also be appropriate practice to consider, and finally to use technology also at the output level of the learning process provided the necessary precautions are taken with regard to its efficacious use. Essentially, the findings suggest that as South Asian educational systems tend towards digitally based education, thoughtfully engaged culturally responsive processes could be formed purposefully and based on the celebration, value and legacies of cultural and linguistic practices in such an educational engagement, with technologically rich learning environments creating more justification behind learning dimensions rather than something equally abstract. Also, the advanced use of mixed quantitative and qualitative measures, predictive analytical measures and qualitative exploration of cultural realities in this study creates a methodological leap in educational technology research. This methodological example, if imitated too closely, changes the basis for future inquiry, thereby shaping the agreed-upon realities and authenticity of the education system. Given the implications of artificial intelligence in classrooms for learners, it is imperative that all research and practice interests focus specifically on the cultural and social realities of educational technology. Only then will educational technology meet the promise of equity, meaningfulness, change and growth, and equitable learning opportunities for all.

**Acknowledgements** We used artificial to proofread this manuscript.

**Conflict of Interest** The authors have no conflict regarding this publication.

**Ethics** The participants consented to this study. This manuscript has not been published anywhere or is under consideration for publication.

## References

Amaya, E. J., Restrepo-Calle, F., & Ramírez-Echeverry, J. J. (2023). Discovering insights in learning analytics through a mixed-methods framework: Application to computer programming education. *Journal of Information Technology Education Research*. https://doi.org/10.28945/5182

Bassi, S., Thapliyal, N., Chaturvedi, N., Radhakrishnan, A., & Arora, M. (2025). Peer education initiatives for promoting adolescent health in South-East Asia Region (SEAR): Findings from desk review and key informant interviews (KIIs). *AJPM Focus*. https://doi.org/10.1016/j.focus.2025.100325

Bruner, J. (1984). Vygotsky's zone of proximal development: The hidden agenda. *New Directions for Child and Adolescent Development*. https://doi.org/10.1002/cd.23219842309

Bui, H. P., & Vo, L.-H. (2024). Interviews: Considerations and techniques. In H. P. Bui (Ed.), *Considerations and techniques for applied linguistics and language education research* (pp. 136–145). IGI Global.

Burch, G. F., Heller, N. A., Burch, J. J., Freed, R., & Steed, S. A. (2015). Student engagement: Developing a conceptual framework and survey instrument. *Journal of Education for Business*. https://doi.org/10.1080/08832323.2015.1019821

Cahyono, J. P., & Mukin, K. K. (2025). Language education, politics and technology in South Asia shaping inclusive societies, identities, and futures. *Asian Ethnicity*. https://doi.org/10.1080/14631369.2025.2474647

Celik, I., Gedrimiene, E., Siklander, S., & Muukkonen, H. (2024). The affordances of artificial intelligence-based tools for supporting 21st-century skills. *Australasian Journal of Educational Technology*. https://doi.org/10.14742/ajet.9069

Daga, S., & Rajan, G. (2023). Investigating linguistic abilities and its relationship with empathy, emotional intelligence and cognitive flexibility. *Psycholinguistics*. https://doi.org/10.31470/2309-1797-2023-33-2-57-89

De Brún, A., Rogers, L., Drury, A., & Gilmore, B. (2021). Evaluation of a formative peer assessment in research methods teaching using an online platform: A mixed methods pre-post study. *Nurse Education Today*, 105–166. https://doi.org/10.1016/j.nedt.2021.105166

Fetters, M. D., & Guetterman, T. C. (2021). Development of a joint display as a mixed analysis. *Routledge eBooks*. https://doi.org/10.4324/9780203729434-23

García-Holgado, A., Vázquez-Ingelmo, A., Shoeibi, N., Therón, R., & García-Peñalvo, F. J. (2024). Enhancing language learning through human-computer interaction and generative AI: LATILL platform. *Lecture Notes in Computer Science*. https://doi.org/10.1007/978-3-031-61691-4_17

Hamadouch, A., & Aoumeur, H. (2024). Teachers' perceptions towards assessing learners' intercultural communicative competence in the EFL classroom in Algeria. *Jurnal Arbitrer, 10*(4), 384–390. https://doi.org/10.25077/ar.10.4.384-390.2023

Hao, Z., Jiang, J., Yu, J., Liu, Z., & Zhang, Y. (2025). Student-AI interaction in an LLM-empowered learning environment: A cluster analysis of engagement profiles. *arXiv (Cornell University)*. https://doi.org/10.48550/arxiv.2503.01694

Henss, L., & Pinquart, M. (2023). Coping with expectation violations in education: The role of optimism bias and need for cognitive closure. *European Journal of Psychology of Education*. https://doi.org/10.1007/s10212-023-00783-5

Hickey, A., & Riddle, S. (2021). Relational pedagogy and the role of informality in renegotiating learning and teaching encounters. *Pedagogy Culture and Society*. https://doi.org/10.1080/14681366.2021.1875261

Hill, M. F. (2020). Student mental health and well-being: Overview and future directions. *Irish Journal of Psychological Medicine*. https://doi.org/10.1017/ipm.2020.110

Hmoud, M., Daher, W., & Ayyoub, A. (2025). From experience to engagement: A mixed methods exploration of learning environments using artificial intelligence and extended reality. *Frontiers in Education*. https://doi.org/10.3389/feduc.2025.1617132

Hockham, D., & Wallis, J. (2023). Operationalising the concept of recontextualization: Leveraging pedagogy and learning strategies in higher education practical courses. *Higher Education*. https://doi.org/10.1007/s10734-023-01136-3

Hu, W., Tian, J., & Li, Y. (2024). Enhancing student engagement in online collaborative writing through a generative AI-based conversational agent. *The Internet and Higher Education.* https://doi.org/10.1016/j.iheduc.2024.100979

Hung, B. P., & Nguyen, L. T. (2022). Scaffolding language learning in the online classroom. In R. Sharma & D. Sharma (Eds.), *New trends and applications in Internet of Things and Big Data analytics* (pp. 46–60). Springer.

Hung, B. P. (2019). A cognitive linguistic approach to teaching English idioms to EFL students: Experimental results. *3L: The Southeast Asian Journal of English Language Studies, 25*(2), 113–126. https://doi.org/10.17576/3L-2019-2502-09

Hung, B. P. (2019). Impacts of cooperative learning: A qualitative study with EFL students and teachers in Vietnamese college. *Issues in Educational Research, 29*(4), 1223–1240. http://www.iier.org.au/iier29/hung.pdf

Isaac, M., Ateeq, M., Hafizh, H., Hu, B., & Shodipo, D. (2023). Leveraging artificial intelligence with zone of proximal development: An ARCS motivational e-learning model. *Leveraging Artificial Intelligence with Zone of Proximal Development.* https://doi.org/10.1109/tale56641.2023.10398344

Kuiper, L., Helthuis, R., Polinder-Bos, H., Lemmens, L., Dollé, M., Slagboom, E., & Van Meurs, J. (2025). Exploring geriatricians' perspectives on aging biomarkers: A reflexive thematic analysis. *Maturitas.* https://doi.org/10.1016/j.maturitas.2025.108601

Lantolf, J. P., Poehner, M. E., & Swain, M. (Eds.). (2018). *The Routledge handbook of sociocultural theory and second language development* (1st ed.). Routledge. https://doi.org/10.4324/9781315624747

Lee, G., Mun, S., Shin, M., & Zhai, X. (2024). Collaborative learning with artificial intelligence speakers. *Science & Education.* https://doi.org/10.1007/s11191-024-00526-y

Lin, M. P., Liu, A. L., Poitras, E., Chang, M., & Chang, D. H. (2024). An exploratory study on the efficacy and inclusivity of AI technologies in diverse learning environments. *Sustainability.* https://doi.org/10.3390/su16208992

Liu, F., & Luo, W. (2025). Exploring the effects of online peer feedback on tertiary students' peer feedback literacy in English writing: Focusing on design elements. *Studies in Educational Evaluation.* https://doi.org/10.1016/j.stueduc.2025.101516

Martin-Beltrán, M., Guzman, N. L., & Chen, P. J. (2017). 'Let's think about it together:' How teachers differentiate discourse to mediate collaboration among linguistically diverse students. *Language Awareness.* https://doi.org/10.1080/09658416.2016.1278221

Omar, A. S., Mgala, M., & Mwakondo, F. (2025). AI-driven visual scaffolding in education: A comprehensive literature review. *International Journal of Research and Scientific Innovation.* https://doi.org/10.51244/ijrsi.2025.12030055

Ornelles, C. (2007). Providing classroom-based intervention to at-risk students to support their academic engagement and interactions with peers. *Preventing School Failure: Alternative Education for Children and Youth.* https://doi.org/10.3200/psfl.51.4.3-12

Pagani, V. (2021). Behind the numbers. A mixed-methods study of the CLASS tool application in Italy. *Journal of Behavioral Education.* https://doi.org/10.7346/sird-012021-p46

Puntambekar, S. (2021). Distributed scaffolding: Scaffolding students in classroom environments. *Educational Psychology Review.* https://doi.org/10.1007/s10648-021-09636-3

Reitman, J. G., Clevenger, C., Beck-White, Q., Howard, A., Rose, S., Elick, J., & D'Mello, S. K. (2023). A multi-theoretic analysis of collaborative discourse: A step towards AI-facilitated student collaborations. *Lecture Notes in Computer Science.* https://doi.org/10.1007/978-3-031-36272-9_47

Sagredo-Gallardo, M., Campos, J. G., Contreras, C. A., & Elias, M. (2025). Desafíos y oportunidades de la inteligencia artificial en el aprendizaje colaborativo: implicancias para la innovación educativa en contextos institucionales. *European Public & Social Innovation Review.* https://doi.org/10.31637/epsir-2026-2211

Sampson, A. E. (2021). Languaging and learning in independent EFL study compared to collaborative pairwork. *International Journal of Educational Research Open*. https://doi.org/10.1016/j.ijedro.2021.100102
Scotland, J. (2022). Exploring the relationship between collaborative dialogue and the learning of L2 form. *International Journal of Educational Research Open*. https://doi.org/10.1016/j.ijedro.2021.100121
Shahini, A. (2025). Emotional dimensions of feedback: How AI and human responses shape ESL learning outcomes. *Ampersand*. https://doi.org/10.1016/j.amper.2025.100235
Smith, M., Koole, M., Adam, T., Traxler, J., & Footring, S. (2024). Methodological insights for decolonising research and EdTech. *Education Sciences*. https://doi.org/10.3390/educsci14060580
Storch, N. (2005). Collaborative writing: Product, process, and students' reflections. *Journal of Second Language Writing*. https://doi.org/10.1016/j.jslw.2005.05.002
Su, Y., Chang, C., Wang, C., & Lai, C. (2022). A study of students' learning perceptions and behaviors in remote STEM programming education. *Frontiers in Psychology*. https://doi.org/10.3389/fpsyg.2022.962984
Tarifa-Rodriguez, A., Virues-Ortega, J., & Calero-Elvira, A. (2024). The behavioral education in social media (BE-SOCIAL) program for postgraduate academic achievement: A randomized controlled trial. *Journal of Behavioral Education*. https://doi.org/10.1007/s10864-024-09545-9
Tin, N. T. & Bui, H. P. (2024). Sampling, generalizability, and reliability. In H. P. Bui, *Applied linguistics and language education research methods: Fundamentals and innovations* (pp103–121). IGI Global.
Van Le, H., Dang, B., Vo, T. P., Nguyen, M. Q., & Nguyen, N. Q. (2025). Anonymous yet accountable: Investigating sociocultural dynamics in peer feedback among Vietnamese EFL students. *Learning Culture and Social Interaction*. https://doi.org/10.1016/j.lcsi.2025.100939
Vygotsky, L. S. (1980). Mind in society. *Teachers and Teaching*. https://doi.org/10.2307/j.ctvjf9vz4
Wang, X., Liu, Q., Pang, H., Tan, S. C., Lei, J., Wallace, M. P., & Li, L. (2022). What matters in AI-supported learning: A study of human-AI interactions in language learning using cluster analysis and epistemic network analysis. *Computers & Education*. https://doi.org/10.1016/j.compedu.2022.104703
Wen, C., Liu, C., Li, C., Chang, M., Chiang, S. F., Lin, H., & Biswas, G. (2024). The learning analytics of computational scientific modeling with self-explanation for subgoals and demonstration scaffolding. *Computers & Education*. https://doi.org/10.1016/j.compedu.2024.105043
Wu, C., & Peng, K. (2025). Development of an artificial intelligence-based text sentiment analysis system for evaluating learning engagement levels in STEAM education. *Applied Sciences*. https://doi.org/10.3390/app15084304
Yan, L.M.-M. (2025b). The effects of generative AI agents and scaffolding on enhancing students' comprehension of visual learning analytics. *Computers & Education*. https://doi.org/10.1016/j.compedu.2025.105322
Yan, L., Martinez-Maldonado, R., Jin, Y., Echeverria, V., Milesi, M., Fan, J., & Gašević, D. (2025). The effects of generative AI agents and scaffolding on enhancing students' comprehension of visual learning analytics. *Computers & Education*. https://doi.org/10.1016/j.compedu.2025.105322
Yan, L. (2025). From passive tool to socio-cognitive teammate: A conceptual framework for agentic AI in human–AI collaborative learning. *arXiv (Cornell University)*. https://doi.org/10.48550/arxiv.2508.14825
Yuen, C. L., & Schlote, N. (2024). Learner experiences of mobile apps and artificial intelligence to support additional language learning in education. *Journal of Educational Technology Systems, 52*(4), 507–525. https://doi.org/10.1177/00472395241238693
Zhang, S., & Yin, Z. (2025). Differences in the collaborative knowledge construction process between high- and low-achievement groups in a collaborative concept mapping task: Evidence

from multiple perspectives. *Thinking Skills and Creativity*. https://doi.org/10.1016/j.tsc.2025.101984

**D. Solomon Paul Raj** serves as an Assistant Professor of English at Sri Krishna Arts and Science College in Coimbatore. He possesses more than a decade of experience in teaching, researching, and leading at an administrative level. His research has led to widely published scholarship in Scopus-indexed journals, focusing on topics related to English language teaching, AI in pedagogy, ecocriticism of English literature, and studies of memory. He has facilitated and presented at conferences (lectures, workshops, and roundtable discussions); with authority, faculty development reports, research based, academic papers, and outreach based conference calls variously constructing a bridge between scholarship and social impact.

**K. Saritha** holds the dual roles of Vice Principal and Assistant Professor in the Department of English et al. Shifa College of Arts and Science, where she is heavily involved in teaching, research, and academic administration. She has presented and published several research papers in reputed journals and conferences. She is completing a part time PhD about gender identity and health humanities in English Literature at Kapagam Academy of Higher Education. In addition, she has organized numerous academic programmes, Faculty Development Programs (FDPs) and skill builds for students. She is interested in English language teaching, communication studies, and gender research as a form of individual and social change, through the education process.

**Hariharasudan Anandhan** is a faculty member at the Department of Language, Culture, and Society, SRM Institute of Science and Technology, India. His scholarly contributions, comprising over 67 research articles related to postmodernism, feminism, religion, education 4.0, education management, and other topics, have been published in esteemed journals. In a crowning achievement, Stanford University and Elsevier have recognized the scholarly endeavors of Dr. A. Hariharasudan by listing me among the world's top 2% of scientists, a testament to my unwavering commitment to intellectual exploration.

**Mari Carmen Gil Ortega** serves as the Director of Teaching within the Department of Education at the University of Bath and is a scholar of immersive learning and applied linguistics. Her portfolio encompasses pedagogical approaches with VR technology, inclusive education, and language teaching research. Maika has contributed as an editor for the Journal of Virtual Worlds Research and is a Senior Fellow of the Higher Education Academy. Her research includes leading the ERASMUS + projects GUINEVERE, and recently working on a VR reconstruction of Bristol Bridge. She has a PhD in Education from the University of Brighton, and an MA in Applied Linguistics from Sussex. Most recently, Maika's talks have focused on AI readiness, teacher preparation, and using immersive tools.

**S. Selvalakshmi** is a Professor in the Department of English at Karpagam Academy of Higher Education, Coimbatore. With a considerable teaching and research background in the areas of English language and literature, her research interests lie in the areas of gender studies, post colonial literature, and cultural studies. She has presented papers and delivered lectures at various national, international conferences, and published several research articles in peer reviewed journals. She had guided several research scholars since then and has also contributed to curriculum development, and academic initiatives at institutional levels. She helps promote critical thinking and interdisciplinary learning and prescribes reading literature in order to encourage the students to find it the most effective medium for intellectual and social growth.

**M. Gopinath** is a faculty member of Mathematics at Sri Krishna Arts and Science College, Coimbatore, Tamil Nadu. He teaches mathematics, provides mentorship for quantitative problem solving and research skills, and is working on current developments in the field of statistics, including data-driven modelling, modern statistical inference, and situating mathematics across interdisciplinary uses with real world problems. He is involved with learner-centred pedagogy and participatory, collaborative projects that advance analytical thinking and practical implications.

# AI Feedback and Learning Assessment

# AI-Mediated Feedback in L2 Writing: Characteristics of Prompts, AI Tools, and Learners' Feedback Use

**Tan Hoai Kim and Hoa Thi Hoang Chau**

**Abstract** Artificial Intelligence (AI) has transformed education. For instance, ChatGPT, a generative AI tool, can provide AI-generated written feedback (AI-WF) to assist second language (L2) learning through signaling inappropriacies in provided texts. The burgeoning interest in using AI tools in writing classrooms shows the rationale for literature on L2 students' experiences with AI feedback in writing. Through an extensive survey of literature, this chapter argues for three essential elements which determine L2 writing quality improvement through AI feedback application. These elements are AI tool(s) in use for feedback, students' prompt(s), and students' revision strategies. As these components are dynamic in nature, it might be essential for students to be trained in using AI and its feedback effectively. The chapter aims to contribute the scant literature on how to use AI feedback in the L2 education context.

**Keywords** Artificial Intelligence · AI-generated written feedback · ChatGPT · L2 writing · Learner autonomy

## 1 Introduction

Feedback is generally defined as teacher or peer comments on learners' inappropriate or unexpected production (Bui & Vo, 2024; Steiss et al., 2024). Well documented as a proliferating interest in L2 education, feedback can be categorized into two main forms: oral form, known as oral feedback, or written form, known as written feedback (Ha & Murray, 2020). Oral feedback and written feedback (WF) are well acknowledged to contribute to L2 acquisition.

T. H. Kim
School of Southern Khmer Language, Culture, Arts and Humanities, Tra Vinh University, Vinh Long Province, Vietnam

H. Thi Hoang Chau (✉)
Tra Vinh University, Vinh Long Province, Vietnam
e-mail: cthhoa@tvu.edu.vn



H. P. Bui and L. P. H. Huong (eds.), *Artificial Intelligence in Education and Academic Research*, Studies in Computational Intelligence 1269,
https://doi.org/10.1007/978-3-032-18392-7_10

Technological advances have transformed the landscape of English language teaching. For instance, technology-mediated learning modality is supposed to enhance affordances and learner autonomy in which learners can personalize their learning (Mekheimer, 2025). One provoking current perspective is using generative artificial intelligence (AI), such as Chat GPT, to automate WF. AI-automated WF has proved to suggest corrections for linguistic inaccuracies, assign scores to written work, and provide qualitative feedback for summative and formative purposes (Yavuz et al., 2025).

Although AI shows its potential in assisting learners in the writing process, several researchers (Liu et al., 2025) have raised concerns about its reliability and depth. One concern persists that the relevance of AI-automated to the contextual factors (Yavuz et al., 2025), although AI may provide text-bound comments (Ranalli et al., 2017). Another concern is that the effectiveness of AI-automated WF depends somewhat on prompting strategies (Kim et al., 2025). suggesting that users should be trained to use prompts as good prompts provide generate high-quality WF. Finally, Lee (2014) believes that the power of WF relies on how learners use feedback to develop further.

As AI-automated WF is an emerging interest in L2 education, there is little literature on how learners perceive and use AI regarding feedback in the writing process. To contribute to the contemporary literature, this conceptual paper argues that the improvement in L2 writing quality is directly or indirectly associated with the prompts used by learners, the AI tool that they use, and their integration of AI-mediated feedback into their writing.

Given the arising suggestion of using AI to improve the quality of writing, this chapter will inform practitioners and researchers about the effective use of AI-automated WF. The topic is, therefore, timely and contributes to the comparatively scare literature on AI-assisted WF with an effort to establish its link to the designated context. After this introduction section, the present chapter will discuss aspects WF provision through the lens of sociocultural theory.

## 2 Written Feedback

Assessing L2 writing used to be confined to error correction. The teacher is supposed to correct errors in students' writing. However, the shifts in L2 writing assessment paradigms have been reconceptualized in which WF is viewed as a multidimensional pedagogical tool that guides learners through both language development and writing competence (Hyland & Hyland, 2006). Influenced by sociocultural theory, EF encompasses guidance on content, organization, genre awareness, communicative purpose, and the overall writing process. This suggests that identifying and correcting grammar inaccuracies included in the assessment and writing processes, depending on proficiency level, objectives, and the context. WF is broadly defined to include all written comments which help students enhance their writing through revision (Ferris, 2006). The effective design of WF enables students to develop their cognitive abilities while gaining independence in their learning and understanding language structures which leads to better linguistic accuracy and improved discourse competencies and successful communication (Aljaafreh & Lantolf, 1994; Lee, 2017).

The assessment frameworks of IELTS writing band descriptors support high-quality WF because they concentrate primarily on content relevance, organization and development, lexical resource and grammatical range and accuracy (IELTS, 2023). The Common Core State Standards (CCSS) also establishes five essential criteria for writing assessment which include claim and development and audience and cohesion and style and conventions. The assessment standards indicate that a test taker can reach the highest score by having small language errors in their manuscript which do not affect reading comprehension. Research indicates that complete WF systems need to handle both basic and advanced writing issues to help students develop their writing abilities. The feedback system needs to deliver its support at the right time while choosing specific content that matches student learning stages and objectives. The writing process receives essential support from WF systems which operate through human teachers or artificial intelligence platforms to develop lasting writing abilities for academic and professional environments. Research indicates that writing feedback leads to better writing quality when students find value in the process and apply it to their work (Ferris, 2006). The WF provider needs to help students detect differences between their actual writing and desired outcomes which enables them to edit their work and learn new language skills (Schmidt, 1990). The combination of metalinguistic explanations with WF enables students to learn and maintain grammatical knowledge better. Students who receive genre-specific feedback develop better understanding of tone and structure and audience needs.

The practice of WF enables students to achieve better linguistic skills while developing their cognitive abilities and self-directed learning skills. The practice of selective WF which limits corrections to avoiding student overload leads to better student confidence and motivation (Lee, 2009). The process of indirect WF makes students identify their mistakes before they correct them, resulting in better critical thinking abilities (Ferris & Roberts, 2001). Students may integrate teacher feedback into their paper before they submit their final draft to the teacher. As feedback can be provided by peers, peer review may benefit learners' critical thinking, language use, and revision in which their interaction can facilitate their language acquisition and offer students constructive feedback which they can integrate into their writing (Min, 2005).

Through the lens of sociocultural theory, feedback serves as a mediator, enabling students to develop intentionality and reciprocity through collaborative work with teachers and students while they solve problems together and transfer learning strategies between tasks and understand their individual learning needs (Lantolf & Thorne, 2006). The delivery of effective feedback may help teachers and educational settings achieve better management and operational efficiency. The implementation of focused WF may assist teachers in decreasing their workload while preserving their teaching effectiveness (Bitchener & Ferris, 2012). Teachers who deliver feedback through scheduled and manageable sessions can fulfill their duties without sacrificing student assistance. The practice of WF creates conditions for teachers and students to develop their professional skills and become more empowered. Writing conferences as an alternative feedback method enables teachers to establish more effective connections with their students. Teachers who test different feedback approaches and conduct classroom-based research develop are likely to stronger professional identities while improving their teaching methods (Lee, 2014).

L2 WF has shifted from basic error correction to a complex method that helps students improve their language skills and thinking abilities and writing ability. The combination of university-school partnerships with teacher education programs will help improve feedback methods. Teachers who join professional learning communities use these spaces to evaluate their feedback methods and enhance their teaching practices (Avalos, 2011). The traditional error-based approach to WF creates negative experiences for teachers and students but strategic implementation transforms it into an effective learning tool that boosts student motivation and writing abilities. The IELTS scoring rubric provides a suitable framework for developing effective WF that should guide assessment processes. The feedback process should expand past basic language accuracy to evaluate content value and structural arrangement and genre understanding and student participation levels. Students who receive holistic feedback develop their writing independence while improving their academic and practical communication abilities. The combination of process-oriented teaching methods with peer feedback and teacher support enables WF to develop students' writing skills. The implementation of WF as an educational tool requires three essential elements which include principled approaches and learner-focused methods and practical applications. The main purpose of WF extends beyond error correction because it needs to help students develop their writing abilities.

Research has established multiple essential elements which make up high-quality WF. The process of giving feedback requires teachers to maintain both specific targets and clear directions. The most effective feedback approach for students involves choosing specific errors which affect their writing performance the most. According to Lee (2017) and Ferris (2006) students who receive numerous corrections become less motivated to learn because their brains struggle to make sense of and utilize the feedback effectively. The process of helping students identify their writing gaps through focused feedback matches the principles of formative assessment which focuses on specific learning targets. Therefore, two essential elements for effective feedback include both precise targeting and consistent application. As the feedback process requires teachers to maintain both precise accuracy and uniform application throughout all student assignments, students may lose trust in feedback when teachers apply different correction methods because of fatigue or absent standardized procedures which makes them less likely to follow the feedback (Lee, 2008a, 2008b).

The effectiveness of feedback depends on how well readers can understand its meaning. The feedback needs to be easy to read and follow a logical structure while avoiding any confusing symbols or unclear statements. Students who have restricted language abilities need clear feedback because they face challenges when trying to understand complex messages (Hyland & Hyland, 2006). The feedback needs to provide meaningful information which helps students understand their mistakes and create improvement plans. The feedback process serves as a diagnostic tool which helps students identify their weaknesses and develop plans for improvement. The feedback process should show students their mistakes and their strengths because this approach helps them become independent writers (Han & Hyland, 2015). The moment when teachers provide feedback to students matters significantly. Students achieve their best results when they receive feedback during the time their minds stay focused on the writing assignment. The connection between errors and learning opportunities becomes weaker when teachers delay their feedback delivery according to Lee (2017). Teachers may face ongoing difficulties when they need to give immediate feedback to their large EFL student population. The feedback process needs to be both actionable and engaging for students. The feedback process should motivate students to edit their work while they are drafting their papers. The instructional value of feedback becomes minimal when students do not take action on the feedback (Voerman et al., 2012). Students who take responsibility for their errors through self-correction develop better cognitive involvement in their work. Students need feedback that shows their achievements in grammar and organization and content development because this helps them stay motivated and build their self-confidence (Hyland & Hyland, 2006). Research-based feedback methods combined with established teaching methods form the foundation for effective feedback practices.

It might be essential that teachers choose their feedback methods based on research evidence while considering student proficiency levels and error types. Indirect WCF is especially beneficial for rule-governed (i.e., treatable) errors as it promotes self-correction and metalinguistic reflection. However, direct comments may be more appropriate for idiosyncratic (untreatable) errors or for learners at the beginner level (Ferris, 2006).

## 3 Dimensions of AI-Mediated WF Effectiveness

While language accuracy remains a central component of WF, a growing body of research argues that effective WF must address a broader range of dimensions in student writing to truly support learning and development. Although scholars have suggested varied criteria for evaluating WF, an extensive survey of the current literature shows the five following dimensions: (1) language form and accuracy, (2) content relevance, (3) text structure, (4) genre, purpose, audience and context, and (5) writing process and development.

Language form and accuracy: Traditionally, feedback in L2 classrooms has focused heavily on grammar, vocabulary, spelling, and punctuation. Teachers often feel obligated to correct surface-level errors due to the prevalence of language issues among learners, particularly in exam-focused environments (Lee, 2017). However, an overemphasis on accuracy can draw students' attention away from requirements for higher-order thinking such as content and organization, which are equally vital for writing proficiency (Hyland & Hyland, 2006). The implementation of effective

WF requires teachers to focus on rule-governed errors which include word forms, article usage and subject-verb agreement while using indirect feedback methods to help students work with linguistic forms.

Content development: Students need feedback that evaluates both the value and clarity and completeness of their generated ideas. Students who receive high-quality content feedback learn to create well-supported arguments and narratives which stay within the assigned topic. Teachers understand the importance of content feedback yet struggle to deliver it because they must handle the extensive work required for language error correction (Lee, 2008a, 2008b). Students benefit from content-focused comments because these comments help them understand their idea strengths and weaknesses and make feedback more effective.

Text organization: The success of writing depends on proper organization through logical order and unified paragraphs and suitable transition connections. The writing elements work together to create better organization and better connection between different parts of the text. Teachers tend to ignore organizational feedback because they focus more on correcting language errors. Students need teacher guidance to create structured texts which present logical progress and clear thinking for better reader understanding.

Genre, purpose, audience, and context: The writing feedback system needs to help students understand different text genres and their intended communication purposes. Students need to demonstrate proper tone and structure and stylistic elements which match their writing assignments. Research indicates that teaching students about different genres enables them to understand what different audiences expect from their writing (Hyland, 2003). Students who write recounts need instruction about creating effective introductions and maintaining logical event sequences.

Writing process and development: Students need feedback at different stages of their writing process which includes drafting and revising and editing. The purpose of feedback extends beyond final evaluation because it functions as a learning tool which supports student development throughout their writing process. Teachers need to establish opportunities for students to assess their work with peers and themselves because this helps students develop self-assessment skills and independence (Ferris, 2006). Students who receive continuous feedback learn to enhance their ideas and identify writing patterns and develop skills that transfer between assignments.

Overall writing competence and learner autonomy: The main objective of WF exists to develop independent writers who can handle their writing tasks independently (Hyland & Hyland, 2006; Lee, 2017). Teachers need to identify learning deficits while teaching students to think about their writing process and help them apply their writing abilities to different situations (Bitchener & Ferris, 2012; Han & Hyland, 2019). Feedback that examines both student strengths and weaknesses lead to increased learner confidence and motivation and better long-term development (Carless & Boud, 2018; Yang & Carless, 2013; Yavuz et al., 2025).

The IELTS scoring rubric serves as an appropriate framework for developing effective WF systems. The feedback system needs to move past language accuracy assessment to evaluate content relevance and organizational structure and genre awareness and student engagement during the writing process. The combination of holistic feedback enables students to develop writing independence while they build deeper cognitive understanding and become proficient academic and real-world communicators. The educational sector has experienced a fundamental transformation because of technology which has led researchers and educators to investigate AI-based learning support systems. Multiple studies from recent years have investigated how AI tools assist students in developing their language abilities through writing instruction (Bitchener & Ferris, 2012). The implementation of AI-based automated writing feedback systems shows promise for developing learner autonomy and individualized learning paths and student engagement (Azhar & Abdullah, 2024). The writing development of EFL students receives support through AI tools which include ChatGPT and Grammarly because these systems generate instant personalized feedback. The adoption of AI tools for writing support in EFL education continues to grow because they provide immediate customized feedback. The effectiveness of AI tools for language development remains uncertain because students in Vietnamese educational settings lack sufficient skills to use these tools correctly. Students demonstrate growing comfort with generative AI yet they struggle to create effective prompts and assess AI feedback quality and apply it effectively to their writing. The current challenge demands educational institutions to develop specific teaching methods which will help students learn AI literacy and feedback literacy and effective AI tool usage for deeper learning and language development.

## 4 Prompting Strategies

Users enter specific instructions through prompts which AI language models use to generate particular responses. The educational writing process depends on prompts as the main connection between students and AI systems because these prompts determine the accuracy and depth of automated feedback (Brown et al., 2020; Liu et al., 2025). AI-assisted learning applications use three main prompting methods which researchers discuss extensively. Zero-shot prompting requires users to provide direct tasks to models without showing examples because models generate responses based on their built-in knowledge and instructions (Brown et al., 2020). The feedback style and focus of the model become more defined through few-shot prompting which presents three examples of correct responses (Brown et al., 2020). The chain-of-thought prompting method requires models to explain their reasoning process step by step which results in more transparent instructional feedback (Wei et al., 2022).

The generated feedback from AI tools depends on the specific prompts users enter even when they use the same tool. The model treats each prompt as an independent communication situation which results in different feedback characteristics (Zhang & Hyland, 2018). The specific wording of a prompt determines what type of feedback the AI system will produce. AI feedback generated from prompts that focus on language accuracy results in basic error correction that focuses on grammatical and lexical aspects. AI feedback that focuses on content and coherence development produces feedback which helps students improve their argument structure and organizational skills. The design of prompts determines how AI feedback functions between correction tools and learning scaffolds according to Chu and Chiu (2025). The quality of AI-generated WF directly depends on the features of the prompt because generic or unhelpful feedback results from poorly designed prompts but targeted prompts produce better results. The ability to create effective prompts requires both technical expertise and pedagogical knowledge (Maloy & Gattupalli, 2024). Students need training to develop their skills in creating and applying prompts because this skill determines the educational value of AI writing tools. Students who lack proper guidance about prompt use will either become too reliant on the tool or fail to understand its output correctly. Students who develop prompt literacy skills acquire AI literacy which helps them understand automated feedback for independent learning (Carless & Boud, 2018).

## 5 AI Tools and Effective WT

Research studies demonstrate that AI-based feedback tools generate feedback quality and focus differ because their design elements include separate architectures and training data and target objectives. The Automated Writing Evaluation (AWE) systems Criterion and Pigai operated with rule-based systems which focused on surface-level linguistic accuracy but modern Large Language Models (LLMs) including ChatGPT and GPT-4 produce detailed feedback that adapts to context (Hung et al., 2022). The success of feedback depends on AI tool development methods which determine which writing elements to focus on and how to represent linguistic and rhetorical and contextual aspects (Stevenson & Phakiti, 2019).

The evaluation of feedback from human or AI sources requires assessment through five essential criteria which include (1) language accuracy and form and (2) content development and (3) text organization and (4) genre and purpose and audience and context and (5) writing process and development (Hyland & Hyland, 2006; Lee, 2017). AI tools excel at detecting grammatical mistakes and stylistic problems yet they struggle to evaluate content coherence and genre understanding and audience needs (Ferris & Roberts, 2001).

AI-mediated feedback needs developers to create systems which fulfill the five essential pedagogical standards. AI tools need to assist students in improving their writing through error correction and guidance for meaning development and organizational improvement and multiple revision stages. According to Hyland (2003) feedback needs to function as a developmental process which enables dialogue between learners and instructors instead of focusing on error correction. Developers together with educators need to create AI-mediated feedback systems which support student writing development through context-specific and pedagogically sound approaches that focus on long-term learning rather than surface-level language correction.

## 6 Learners' Use of AI-Automated Feedback

The integration of AI-automated feedback with traditional human feedback stands as a leading educational innovation in writing instruction. The two feedback systems operate through different mechanisms which produce distinct results in terms of content delivery and emotional impact. Human feedback exists within social environments because it contains emotional elements and teachers understand student cultural and linguistic backgrounds (Lee, 2017; Maloy & Gattupalli, 2024). Teachers use their classroom relationships and learner affective needs to determine their feedback delivery methods and tone and level of encouragement. AI feedback production depends on standardized training data which enables fast and consistent delivery of feedback that remains emotionless (Wilson & Roscoe, 2020). The neutral delivery of AI feedback reduces emotional risks and biases yet creates challenges for students to feel connected to the feedback process (Hyland & Hyland, 2006; Voerman et al., 2012).

AI systems demonstrate exceptional ability to detect basic writing errors which include grammar mistakes and spelling errors and word selection and sentence structure problems (Stevenson & Phakiti, 2019; Zhang & Hyland, 2018). AI systems lack the ability to provide high-level feedback about argumentation and organization and logic and stylistic elements which peer and teacher feedback excels at. Human feedback enables students to develop their cognitive abilities through reflective activities which lead to metacognitive development when teachers engage in dialogic interactions (Han & Hyland, 2015). AI feedback provides immediate results at scale which benefits large classrooms because teachers cannot provide individualized feedback due to time constraints and emotional exhaustion (Lee, 2008a, 2008b). The flexible nature of teacher feedback allows teachers to adjust their feedback based on student development needs.

AI feedback requires students to evaluate its suggestions through reflective analysis by considering genre requirements and rhetorical purposes and local context needs (Lee, 2017). Students need to validate AI suggestions through teacher or peer feedback when they want to enhance their logic and creativity and idea development skills (Hyland & Hyland, 2006). Students need to receive higher-order guidance during their multiple draft revisions (Wilson & Roscoe, 2020). AI tools function best as error detection systems but they need human feedback to direct students toward effective revision practices (Stevenson & Phakiti, 2019). AI tools function best when they work with human feedback to provide students with complete revision guidance (Stevenson & Phakiti, 2019). Students who learn to evaluate AI feedback develop critical thinking skills and self-regulation abilities through the process of deciding which feedback to accept or modify (Mekheimer, 2025). The process of reflection about AI feedback helps students develop their ability to use AI tools independently. Students may develop critical thinking abilities and metacognition to benefit from AI tools yet these skills develop better through human teacher support (Han & Hyland, 2015; Voerman et al., 2012). The most effective writing development method for

EFL classrooms today involves using AI tools with their precise feedback capabilities together with teacher and peer feedback that provides contextual understanding and emotional support.

The process of writing development starts with receiving feedback but students need to apply this feedback to their subsequent drafts for actual improvement (Bitchener & Ferris, 2012; Han & Hyland, 2015). The quality of student writing before submission depends on their ability to effectively use AI-mediated feedback (Liu et al., 2025). Students need to assess which feedback suggestions match their task requirements and then connect them to their rhetorical objectives and disciplinary standards (Lee, 2017). Students who use AI feedback strategically while maintaining critical thinking abilities will achieve better results in five vital writing areas which include language accuracy and content development and text organization and genre and purpose and audience and context and writing process development (Hyland & Hyland, 2006; Lee, 2017). The essential writing competencies that students develop through AI feedback integration include these five core areas.Students who engage with AI feedback develop methods to prevent future writing mistakes while they improve their current drafts (Han & Hyland, 2019; Zhang & Hyland, 2018). Students develop lasting knowledge about their writing flaws through repeated analysis of grammatical mistakes and unclear arguments and poor transition connections (Chu & Chiu, 2025). AI feedback serves as a metacognitive instrument which enables students to track their writing development through multiple drafts while learning efficient revision techniques (Wilson & Roscoe, 2020). Students who fail to evaluate AI feedback critically will struggle to make independent decisions about language and rhetorical choices (Azhar & Abdullah, 2024) Students need instruction to use AI feedback effectively through reflective practice which views AI feedback as a learning tool instead of a final editing solution (Han & Hyland, 2015). When used in this way, AI tools and human feedback complement each other: AI provides efficiency and linguistic precision, while teachers and peers offer contextual understanding and affective support (Hyland & Hyland, 2006). A hybrid feedback model that combines both forms may thus represent the most pedagogically sound approach for promoting critical engagement, autonomy, and sustained improvement in L2 writing (Han & Hyland, 2015; Stevenson & Phakiti, 2019).

## 7 AI-Mediated WF Under the Lens of Sociocultural Theory

Sociocultural theory has been considered appropriate for analyzing AI-mediated WF because it describes human–computer interaction. Accordingly, it particularly explains how communication enables people to construct their knowledge base (Vygotsky, 1978). The use of computer-mediated communication possibly offers opportunities for communication between people from different cultures (Hung et al., 2022). The core concept in sociocultural theory describes how humans use artifacts to help them transform their knowledge. Research indicates that mediation exists in three distinct forms which include human assistance and self-directed learning and artifact-based support. The process of knowledge exchange between people constitutes the primary function of mediation by others in language education. Vygotsky (1978) initially explained this process through his description of how experts transform novice speakers' understanding during early SCT development. Research conducted in recent years about language education demonstrates that students can transform each other's language through social dialogue (Storch, 2013). The process of knowledge co-creation between learners needs them to use both their thinking abilities and their ability to think about their thinking (Lantolf & Thorne, 2006; Swain, 2006; Zhang & Hyland, 2018). Learners who use self-mediation techniques perform their own language development work. Through private speech learners can process their experiences and acquired knowledge which enables their knowledge to advance through internal mental operations (Lantolf, 2000; Vygotsky, 1978; Poehner & Infante, 2018).

The third type is mediation with artefacts. Teachers in educational settings use tasks (symbolic) and activities (physical) to help students develop their thinking abilities (Aljaafreh & Lantolf, 1994; Lantolf & Thorne, 2006; Wertsch, 1991). Research indicates that knowledge development emerges from the combination of these three mediation approaches (Lantolf, 2010; Swain et al., 2015; van Compernolle, 2019). The process of self-mediation starts after learners receive input from others and work with artifacts because these sources provide the necessary material for their internal processing (Lantolf, 2013). Teachers use mediation through WF to correct student language in their L2 writing assignments. The teacher uses language and task artifacts to indicate and modify student language errors (Aljaafreh & Lantolf, 1994; Han & Hyland, 2019; Nassaji & Swain, 2000). The effectiveness of WF depends on student attention to the feedback because students need to review their mistakes after receiving notification about their errors (Bitchener & Ferris, 2012; Lee, 2020; Storch & Wigglesworth, 2010). The teacher uses WF to indicate errors to students who then need to dedicate time to error correction for future avoidance of repeated mistakes which leads to language development (Han & Hyland, 2015; Yavuz et al., 2025).

## 8 Conclusion

The fundamental purpose of WF in writing education remains essential but artificial intelligence has reshaped its delivery methods and structure. Teacher feedback stands as the essential tool for complete and context-specific assistance because it incorporates pedagogical expertise and student cultural and linguistic backgrounds. Students who receive peer feedback learn to work together while evaluating their work develop active participation in feedback activities and collaborative meaning construction skills. AI-mediated feedback operates as a third feedback system which produces fast results at big scales through precise language analysis to support human teacher work. AI tools produce feedback with different quality levels because their design components and training materials and operational goals differ (Zhang & Hyland,

2018). AI feedback systems produce feedback with different quality levels because their design elements and training materials and operational targets have different characteristics. AI feedback systems produce feedback with different quality levels because their design elements and training materials and operational targets have different characteristics. The assessment of human-made and AI-generated feedback needs evaluation through five fundamental criteria which include language form and accuracy and content development and text organization and genre and purpose and audience and context and writing process and development (Hyland & Hyland, 2006; Lee, 2017). Students who receive feedback that meets these standards can make meaningful revisions which results in improved writing development. The quality of AI feedback output depends directly on the specific input prompts users enter when using the system. The quality of AI feedback depends on prompt specificity because specific prompts generate superior results than general or vague prompts (Liu et al., 2025). The educational value of AI-based feedback emerges from two aspects: the automated comments it generates and how students handle these comments during their revision process. Students who analyze their feedback choices between accepting and modifying and rejecting develop their self-regulated learning abilities and metacognitive skills. Students require training to use prompts correctly and assess AI-generated feedback critically because this helps them develop writing skills and teaching independence and technology-based learning that supports human needs.

**Acknowledgements** We used ChatGPT to proofread this manuscript.

**Conflict of Interest:** The authors have no conflict regarding this publication.

**Ethics** This manuscript is original. It has not been published anywhere or is under consideration for publication.

## References

Aljaafreh, A., & Lantolf, J. P. (1994). Negative feedback as regulation and second language learning in the zone of proximal development. *Modern Language Journal, 78*(4), 465–483. https://doi.org/10.1111/j.1540-4781.1994.tb02064.x

Avalos, B. (2011). Teacher professional development in teaching and teacher education over ten years. *Teaching and Teacher Education, 27*(1), 10–20. https://doi.org/10.1016/j.tate.2010.08.007

Azhar, A., & Abdullah, A. (2024). Artificial intelligence (AI) in language learning autonomy (LLA): A systematic literature review uncovering learning autonomy. *International Journal for Multidisciplinary Research, 6*(6), 1–17. https://doi.org/10.36948/ijfmr.2024.v06i06.31045

Bitchener, J., & Ferris, D. R. (2012). Written corrective feedback in second language acquisition and writing. *Routledge*. https://doi.org/10.4324/9780203832400

Brown, T. B., Mann, B., Ryder, N., Subbiah, M., Kaplan, J., Dhariwal, P., & Amodei, D. (2020). Language models are few-shot learners. *Advances in Neural Information Processing Systems, 33*, 1877–1901. https://doi.org/10.48550/arXiv.2005.14165

Bui, H. P., & Vo, T. H. (2024). Revisiting oral corrective feedback in second language acquisition: Existing debates and directions for further studies. In T. C. Bang, C. H. Nguyen, & H. P. Bui (Eds.), *Exploring contemporary English language education practices* (pp. 97–115). IGI Global.

Carless, D., & Boud, D. (2018). The development of student feedback literacy: Enabling uptake of feedback. *Assessment & Evaluation in Higher Education, 43*(8), 1315–1325. https://doi.org/10.1080/02602938.2018.1463354

Chu, K. Y., & Chiu, P. M. H. (2025). Prompt design and AI response quality in University Students' academic learning tasks. *Proceedings of the Association for Information Science and Technology, 62*(1), 1402–1404. https://doi.org/10.1002/pra2.1417

Ferris, D. R., & Roberts, B. (2001). Error feedback in L2 writing classes: How explicit does it need to be? *Journal of Second Language Writing, 10*(3), 161–184. https://doi.org/10.1016/S1060-3743(01)00039-X

Ferris, D. R. (2006). Does error feedback help student writers? New evidence on the short- and long-term effects of written error correction. In K. Hyland & F. Hyland (Eds.), *Feedback in second language writing: Contexts and issues* (pp. 81–104). Cambridge University Press. https://doi.org/10.1017/CBO9781139524742.007

Ha, P. L., & Murray, D. E. (2020). Silence as right, choice, resistance and strategy among East Asian students in transnational English classrooms. *Discourse, 41*(3), 336–352. https://doi.org/10.1080/01596306.2012.745733

Han, J., & Hyland, F. (2015). Exploring learner engagement with written corrective feedback in a Chinese tertiary EFL classroom. *Journal of Second Language Writing, 30*, 31–44. https://doi.org/10.1016/j.jslw.2015.08.002

Han, J., & Hyland, F. (2019). Learner engagement with written corrective feedback: A sociocognitive perspective. *Applied Linguistics, 40*(2), 385–412. https://doi.org/10.1017/9781108635547.015

Hung, B. P., Pham, A. T. D., & Purohit, P. (2022). Computer mediated communication in second language education. In *New trends and applications in IoT and big data analytics* (pp. 45–60). https://doi.org/10.1007/978-3-030-99329-0_4

Hyland, K. (2003). Second language writing. *Cambridge University Press*. https://doi.org/10.1017/CBO9780511667251

Hyland, K., & Hyland, F. (2006). Feedback in second language writing: Contexts and issues. *Cambridge University Press*. https://doi.org/10.1017/CBO9781139524742

IELTS. (2023). *IELTS writing band descriptors (public version)*. British Council.

Kim, J., Yu, S., Lee, S. S., & Detrick, R. (2025). Students' prompt patterns and its effects in AI-assisted academic writing. *Journal of Research on Technology in Education*, 1–18. https://doi.org/10.1080/15391523.2025.2456043

Lantolf, J. P. (Ed.). (2000). *Sociocultural theory and second language learning*. Oxford University Press.

Lantolf, J. P., & Thorne, S. L. (2006). Sociocultural theory and the genesis of second language development. *Oxford University Press*. https://doi.org/10.1093/applin/amm027

Lantolf, J. P. (2010). *Sociocultural theory and the pedagogical imperative*. https://doi.org/10.1093/oxfordhb/9780195384253.013.0011

Lantolf, J. P. (2013). Sociocultural theory: A dialectical approach to L2 research. In C. A. Chapelle (Ed.), *The Routledge handbook of second language acquisition* (pp. 57–72). Routledge. https://doi.org/10.4324/9780203808184

Lee, I. (2008a). Student reactions to teacher feedback in two Hong Kong secondary classrooms. *Journal of Second Language Writing, 17*(3), 144–164. https://doi.org/10.1016/j.jslw.2007.12.001

Lee, I. (2008b). Ten mismatches between teachers' beliefs and written feedback practice. *ELT Journal, 63*(1), 13–22. https://doi.org/10.1093/elt/ccn010

Lee, I. (2014). Revisiting teacher feedback in EFL writing from sociocultural perspectives. *TESOL Quarterly, 48*(1), 201–213. https://doi.org/10.1002/tesq.153

Lee, I. (2017). Classroom writing assessment and feedback in L2 school contexts. *Springer*. https://doi.org/10.1007/978-981-10-3924-9

Liu, M., Zhang, L. J., & Neufeld, T. J. (2025). Chinese EFL learners' GenAI literacy. *Innovation in Language Learning and Teaching* 1–26. https://doi.org/10.1080/17501229.2025.2549754

Maloy, R.W., & Gattupalli, S. (2024). Prompt literacy. *EdTechnica*, 211–216. https://doi.org/10.59668/371.14442

Mekheimer, M. (2025). Generative AI-assisted feedback and EFL writing. *Discover Education, 4*(1), 170. https://doi.org/10.1007/s44217-025-00602-7

Min, H. T. (2005). Training students to become successful peer reviewers. *System, 33*(2), 293–308. https://doi.org/10.1016/j.system.2004.11.003

Nassaji, H., & Swain, M. (2000). A Vygotskian perspective on corrective feedback. *Language Awareness, 9*(1), 34–51. https://doi.org/10.1080/09658410008667135

Ranalli, J., Link, S., & Chukharev-Hudilainen, E. (2017). Automated writing evaluation. *Educational Psychology, 37* (1), 8–25. https://doi.org/10.1080/01443410.2015.1136407

Schmidt, R. (1990). The role of consciousness in second language learning. *Applied Linguistics, 11*(2), 129–158. https://doi.org/10.1093/applin/11.2.129

Steiss, J., Tate, T., Graham, S., Cruz, J., Hebert, M., Wang, J., & Olson, C. B. (2024). Comparing the quality of human and ChatGPT feedback. *Learning and Instruction, 91*, Article 101894. https://doi.org/10.1016/j.learninstruc.2024.101894

Stevenson, M., & Phakiti, A. (2019). Automated feedback and second language writing. *Feedback in Second Language Writing*, 125–142. https://doi.org/10.1017/9781108635547.009

Storch, N., & Wigglesworth, G. (2010). Learners' processing and retention of corrective feedback. *Studies in Second Language Acquisition, 32*(2), 303–334. https://doi.org/10.1017/S0272263109990532

Storch, N. (2013). *Collaborative writing in L2 classrooms* (vol. 31). Multilingual Matters. https://doi.org/10.21832/9781847699954

Swain, M. (2006). Languaging, agency and collaboration. In H. Byrnes (Ed.), *Advanced language learning* (pp. 95–108). Continuum.

Swain, M., Kinnear, P., & Steinman, L. (2015). *Sociocultural theory in second language education* (2nd ed.). Multilingual Matters. https://doi.org/10.21832/9781783093182

Van Compernolle, R. A. (2019). Constructing a second language sociolinguistic repertoire. *Applied Linguistics, 40*(6), 871–893. https://doi.org/10.1093/applin/amy033

Voerman, L., Meijer, P. C., Korthagen, F. A., & Simons, R. J. (2012). Types and frequencies of feedback interventions. *Teaching and Teacher Education, 28*(8), 1107–1115. https://doi.org/10.1016/j.tate.2012.06.006

Vygotsky, L.S. (1978). *Mind in society: The development of higher psychological processes*. Harvard University Press.

Wei, J., Wang, X., Schuurmans, D., Bosma, M., Ichter, B., Xia, F., Chi, E.H., Le, Q.V., & Zhou, D. (2022). Chain of thought prompting. *NeurIPS, 35*, 24824–24836. https://doi.org/10.48550/arXiv.2201.11903

Wertsch, J. V. (1991). *Voices of the mind: A sociocultural approach to mediated action*. Harvard University Press.

Wilson, J., & Roscoe, R. D. (2020). Automated writing evaluation and feedback. *Journal of Educational Computing Research, 58*(1), 87–125. https://doi.org/10.1177/0735633119830764

Yang, M., & Carless, D. (2013). The feedback triangle. *Teaching in Higher Education, 18*(3), 285–297.

Yavuz, F., Aksoy, B., & Kılınç, B. (2025). AI-powered written feedback for EFL learners. *Asian EFL Journal, 27*(1), 21–39. https://doi.org/10.1080/13562517.2012.719154

Zhang, P., & Hyland, K. (2018). Student engagement with teacher and automated feedback. *Assessing Writing, 37*, 13–24. https://doi.org/10.1016/j.asw.2018.02.004

**Tan Hoai Kim** is serving as a visiting lecturer at Tra Vinh University, Vietnam. His academic work centers on teaching English to students who learn it as their second language and particularly investigates how technology supports learning and how teachers can use generative AI to improve language education. His research goal involves establishing competence as an essential element for English curriculum development and educational policy creation.

**Hoa Thi Hoang Chau** is the Director of Department of International Affairs, Tra Vinh University, Vietnam. She earned her Ph.D. in TESOL from Hue University and her M.Ed. and B.Ed. in TESOL from Can Tho University. Her research focuses on English language education and teacher training and EMI and online learning and cultural integration in EFL classrooms to develop intercultural communicative competence.

# Ethical, Pedagogical, and Research Implications of AI Use in Education and Academic Research

**Le Pham Hoai Huong and Thuy Thanh Dao**

**Abstract** The evolution of AI has transformed education and academic research. It powers several academic tasks, such as teaching, learning, implementing research, and research writing. For example, generative AI could produce human-like outputs in response to user input, such as providing constructive feedback, outlines key research interests in a particular area, and summarizing an input text. However, researchers and practitioners have raised concerns about possible risks of AI. Central to these concerns are ethics, accountability, and integrity. Acknowledging the current concepts about AI use, this chapter contributes to the still limited literature on ethical use of AI. The chapter first provides an overall landscape of AI use in education and academic research, showing both potentials and achievements of AI before raising ethical concerns about responsible AI use. Subsequently, we outlines issues regarding the AI use in the academic world, arguing for the roles of policy and guidelines. Finally, we outlines issues that the future generations of AI should respond to.

**Keywords** Academic research · Accountability · Artificial intelligence · Education · Ethics

## 1 Introduction

Advancements in AI technologies have revolutionized many disciplines, including education and academic research. Generative AI using large language models, such as ChatGPT, has demonstrated considerable potential in generating coherent texts (Bernabei et al., 2023; Cotton et al., 2023; Pepin et al., 2025). However, scholars (e.g. Ali et al., 2024; Nadim & Fucio, 2025; Pikhart & Al-Obaydi, 2025) have

L. P. H. Huong (✉)
Hue University of Foreign Languages and International Studies, Hue University, Hue, Vietnam
e-mail: lphhuong@hueuni.edu.vn

T. T. Dao
Faculty of Social Science, Hong Duc University, Thanh Hoa, Vietnam
e-mail: daothanhthuy@hdu.edu.vn



H. P. Bui and L. P. H. Huong (eds.), *Artificial Intelligence in Education and Academic Research*, Studies in Computational Intelligence 1269,
https://doi.org/10.1007/978-3-032-18392-7_11

raised concerns about several aspects of AI use in education and academic research. Although these experts are from different cultures and belong to different disciplines, they all have proposed the importance for internationally and nationally shared guidelines for AI use.

The traditional definition of plagiarism is mainly confined to similarity, i.e. copying and pasting without acknowledging the source. The development of generative AI, enabling the generation of content with human input, has provided opportunities for copying AI-generated contents without any similarity rate. Fortunately, AI detectors may identify such contents (Cotton et al., 2023). This practice should be regarded as a form of academic dishonesty, causing unfairness and violating integrity in the academic world. However, different AI detectors may yield different results, suggesting that AI detectors may provide false positive and false negative levels of AI-generated contents (Weber-Wulff et al., 2023; Weichert & Dimobi, 2024).

Awareness of AI as a tool reminds us of its dual nature, facilitating and disrupting users when it is employed without ethical considerations. This chapter discusses the ethical concerns of AI use and implications for its adoption in education and academic inquiry. It consists of four main sections in addition to this introduction. In each section, we discuss the current perspectives on AI use, supported by empirical findings from recent research before guidelines to imrpove the academic world. To conclude, we discuss the future directions for AI development and AI use; we argue that AI will develop further to respond to the current shortcomings of AI. It may be essential for educational institutions to develop standardized guidelines for AI use and make them accessible to stakeholders.

## 2 Ethical Issues in AI Use in Education and Research

Ethical use of AI has been raised regarding matters of integrity and authorship (Perkins, 2023). As for integrity, the Tertiary Education Quality and Standards Agency (TEQSA, 2022, para. 1) defines it as, "the expectation that teachers, students, researchers and all members of the academic community act with: honesty, trust, fairness, respect and responsibility". Perkins (2023) summarizes academic integrity violation as academic misconduct that constitutes a breach of academic integrity principles, whether this is intentional or not (p. 8). Other scholars (e.g. Bin-Nashwan et al., 2023; Ramdani, 2018) have also argued that academic integrity can be reflected through research and academic tasks with fairness and coherence.

Despite the fact that the landscape of AI integration has been rapidly changing, its acceptance and application are highly contested (Afroogh et al., 2024; Kelly et al., 2023). For instance, Kasneci et al. (2023) described the mixed integration of AI across higher education institutions. While some institutions tend to confine AI use, others have integrated it into their curricular practices with conditions and ethical guidance. This has raised the issue of AI literacy which Tour et al. (2025) conceptualize as the competence to understand, evaluate, and document AI use in learning.

For AI tools to be a supporting artifact used ethically, there must be standardized policies within a country and across contexts so that students and academics are aware of its use (Funa & Gabay, 2025; Nguyen et al., 2023; Turková, 2025). In particular, statements about what is permitted and/or considered to be misconduct should be exemplified. Regarding students' academic assignments, Tour et al. (2025) proposed the need to include a part of students' acknowledgement of their AI use, submitting both AI-assisted drafts and their own revised versions for accountability and reflections.

To ensure AI tools function effectively as learning resources, there must be ethical guidelines for them to be used responsibly in education (Burger et al., 2023; Johnson & Davis, 2024; Kangwa et al., 2025). According to these authors, higher education institutions must provide frameworks to support equitable and sustainable use of AI for academic integrity across educational settings. This is especially critical when assessing students' work and performance (Barus et al., 2025; Chan, 2023). Students must be informed of the ethical principles and guidelines governing AI usage in their study (Marín et al., 2025; Temper et al., 2025). For instance, Eden et al. (2024) propose that institutions must establish clear lines of responsibility and accountability for the design, implementation, and biases of AI systems. Also, it is necessary to raise teachers and students' awareness of the mechanisms for monitoring and evaluating the AI tools for fair and responsible use.

Overall, the ethical integration of AI into education and academic research requires coordiated institutional effort in policy development, AI literacy, transparent assessment practices, continuous understanding. Clear guidelines, shared responsibilities, and systematic governance are essential to ensure academic integrity and quality of education and research. Ethical use of AI thus should take into account the following issues:

- Standardized policies for AI use: Define acceptable and unacceptable AI practices with explicit examples.
- AI literacy development: Provide programs that support Researchers, professionals, teachers, and students to understand, evaluate, and apply AI tools ethically and responsibly.
- Transparency and accountability in academic work: Include statements/declarations of AI use and submit AI-assisted drafts.
- Equitable use of AI tool: Make sure AI use does not undermine critical thinking and academic integrity.
- Institutional governance: Manage AI biases and ensure fair assessment practices, and maintain academic integrity standards.

## 3 AI Use and Pedagogy

Research has increasingly highlighted the contributions of AI in pedagogy. The rise of AI has enhanced efficiency in education as it facilitates personalized autonomous learning (Wang et al., 2024), educational management (Kafar, 2025; Machkour &

Abriane, 2025), student enrollment (AbuDaabes et al., 2025), and pedagogical processes (Al-Zahrani & Alasmari, 2024; Tan et al., 2025). Baker and Smith (2019) proposed a framework for using AI tools in education. For learners, AI tools support them to improve in learning subject matters via specific practice patterns and reflective feedback. ChatGPT, a widely used generative AI tool, is known for providing immediate feedback based on the learner's input and personalizing their learning. A meta-analysis conducted by Dong and colleagues (2025) showed that AI has significant positive impacts on learners' achievements. As generative AI enables the creation of content, it has the potential to aid learners in seeking knowledge in case of the absence of teachers. Several studies (e.g. Godwin et al., 2015; Steinmayr et al., 2016) making use of academic scores, questionnaires, and tests have shown that responsible use of AI is a key indicator of educational success. For teachers, AI-powered tools help them to reduce workload via the use of automated processes in grading, feedback mechanisms, classroom management, administrative issues. For instance, *GradeScanner* can automatically grade multiple-choice tests. For institutions, algorithms are capable of providing processed data mainly for institutional administrators or stakeholders. AI-powered software can process students' transcripts. Tang (2024) also proposed various ways that AI has been used in education by providing smart contents with AI assistants, intelligent tutoring systems, and providing feedback, which make it appealing in education.

From a sociocultural theory perspective, AI-powered tools have the potentials to co-construct knowledge with learners, making it considered to the "more knowledgeable others" as in sociocultural theory (Huong & Bui, 2021). AI tools have the capacity to scaffold just beyond the learner's current capabilities (Kohnke, 2025; Schmidt & Strasser, 2022). They are also tools for teachers' professional development, co-constructing knowledge and plans to adapt teaching strategies and personalized learning experience (Al-Zahrani & Alasmari, 2024).

However, there are debates over the use of AI tools in education. Al-Zahrani and Alasmari (2024), for example, pointed out that AI use might discourage the development of higher-order and critical thinking. These authors are also concerned about balancing AI's and teachers' roles in engaging in communication and interaction for meaningful learning. At the macro-level, effective application of AI tools require advanced infrastructures, instructors with AI literacy which are especially challenging in developing countries (Tang, 2024). Furthermore, to use the tools well, teachers need to master the AI skills and need to be trained. AI algorithms are complicated and hard to understand; thus, when they are used to assess or to detect plagiarism, they outcomes may be unfair and lack of transparency (Celik et al., 2022).

Overall, the use of AI in education offers significant opportunities for students, teachers, and institutions. However, its effective integration requires considerations of challenges faced by users. AI tools can scaffold learners, provide personalized feedback, and streamline administrative work, overreliance on it may hinder higher-order thinking, reduce meaningful teacher-student interaction, and introduce risks due to bias in algorithms. Thus, higher education institutions must ensure that academic integrity policies are explicit in how these tools may be used for education quality. In short, the pedagogical implications of AI adoption are.

- AI enhances efficiency in teaching, learning, and administrative tasks.
- AI can serve as a cultural scaffold in learning and professional development.
- Overreliance on AI may hinder critical thinking and reduce teacher-student interaction.
- Effective use of AI requires strong infrastructures and AI literacy.
- Bias and transparency, and fairness must be addressed in clear institutional policies.

## 4 AI Use in Academic Research

In academic research, like in education, AI works as a transformative tool for researchers since it offers unprecedented capabilities in analyzing data, generating hypothesis, and synthesizing knowledge (Shah et al., 2025). It can work as a catalyst for both methodological innovation and more scholarly insights (Pal, 2023), an active participant in the research process (Burger et al., 2023), and peer reviewer in publications (Kousha & Thelwall, 2023).

However, a range of concerns has been raised regarding bias, transparency, authorship, and data privacy in studies. More specifically, Shah et al. (2025) have raised eight concerns about the rise of AI in academic research. The first is algorithmic bias and scientific fairness due to predictive research patterns and skewed data representation and algorithmic decision-making creating biases related to race, gender, or socioeconomic status. Next is transparency and explainability which are questionable since big data has opaque characteristtics, leading to barriers to reproducibility usually required in academic research. The third concern is related to authorship, accountability, and the intellectual integrity because it is quite challenging to decide intellectual contribution in studies with AI use, which leads to ethical questions and ambiguity in publication and patenting. Another issue is privacy and data protection since AI systems develop on large datasets raising the question of long-term data governance. In this initial stage, inconsistent regulations for AI indicates that research communities must develop specificic requirements for research implementation. Because of the interdisciplinary landscape in research, there should also be ethical codes for cross-research fields requiring collaboration among scientists for holistic insights into AI use. Lastly, Shah et al. (2025) argue that there must be translational ethics to reduce limitations of abstract ethical frameworks. Other concerns such as plagiarism and academic dishonesty (Bin-Nashwan et al., 2023) and ethical, epistemological, and ontological issues (Butson & Spronken-Smith, 2024) also require much attention.

In short, AI has become an indispensable part in academic research for its roles in enhancing methodological innovation, data analysis, and publication. However, its use is associated with various challenges for ethical, fair, and responsible use, which requires regulatory standards, research integrity guidelines, and collaboration across disciplines. In short, AI adoption in research provides:

- AI enhances research productivity via methodological innovation, data systhesis, and result publication.
- AI introduces risks of algorithmic biases and unfairness.
- Transparency and authorship become problematic because of intellectual responsibility and accountability.
- Privacy and data protection concern due to large datasets and unclear long-term data governance.
- Clear guidelines and ethical frameworks are needed in cross-disciplinary research.

## 5 Future Directions

AI has become a defining term in this era in educational and research practices. Future adoption of AI in these fields promises increasingly powerful tools, while simultaneously introducing new risks. Godwin-Jones (2022) highlights the possible co-creation role that AI tools and humans may have to play in future educational scenarios. It is likely that the future integration of AI tools into the classroom environment and research contexts is highly likely, and therefore there must be future policy development giving instructions on its use. AI tools, if used, should be stated clearly and made transparent by all parties.

At the macro level, it is essential to have formal ethical frameworks providing guidelines for scientific inquiry. Thus, "institutions should adopt proactive ethical review systems that integrate AI-specific criteria, including transparency, data privacy, algorithmic fairness, and explainability" (Shah et al., 2025, p. 187). AI use reference frameworks should be designed by multidisciplinary ethics boards for more holistic governance and diverse disciplinary insights.

At the institutional level, ethics-oriented AI training related to research integrity should be prioritized. Institutions must continue to train professionals, teachers, and students for AI tools use efficiency and responsibility. AI literacy development for online environment in learning and researching in multiple disciplines and contexts will be essential in preparing stakeholders for future digital academic environments.

Since research dissemination is ever increasing, publishers, editors, sponsors, as well as academic institutions must concert efforts to provide ethics for authorship, for example, AI as co-authors in case a researcher plans to involve AI language models in writing (Bin-Nashwan et al., 2023). Specific guidelines for AI tools to be acknowledged as research contributors or co-authors must be provided.

Given all the benefits that AI tools in education and research, more investigation, analysis, and evidence, showing the efficacy of AI (Khatun et al., 2024) is recommended to understand their efficacy, limitations, and long-term impacts. Besides, the development of policies and strategies for AI adoption will enable educational institutions to better student-centered online learning environments. More research into AI employment in education and academic inquiry is needed "to provide evidence-based guidelines and support for AI applications in education" (Zhang & Aslan,

2021, p. 8). Evidently, AI rise has given mankind skepticism, doubts, and fears, but when used ethically and responsibly, it continues to open up new possibilities for innovations in these fields.

## References


AbuDaabes, A. S., Selim, H. M., & Hijazi, R. (2025). From clicks to campus: How AI-powered digital marketing shapes student enrollment decisions. *Asia-Pacific Journal of Business Administration (Advance Online Publication).* https://doi.org/10.1108/APJBA-05-2025-0398

Afroogh, S., Akbari, A., Malone, E., Kargar, M., & Alambeigi, H. (2024). Trust in AI: Progress, challenges, and future directions. *Humanities and Social Sciences Communications, 11*, 1568. https://doi.org/10.1057/s41599-024-04044-8

Ali, O., Murray, P. A., Momin, M., Dwivedi, Y. K., & Malik, T. (2024). The effects of artificial intelligence applications in educational settings: Challenges and strategies. *Technological Forecasting and Social Change, 199*, Article 123076. https://doi.org/10.1016/j.techfore.2023.123076

Al-Zahrani, A. M., & Alasmari, T. M. (2024). Exploring the impact of artificial intelligence on higher education: The dynamics of ethical, social, and educational implications. *Humanit Soc Sci Commun, 11*, 1–12. https://doi.org/10.1057/s41599-024-03432-4

Baker, T., & Smith, L. (2019). Educ-AI-tion rebooted? Exploring the future of artificial intelligence in schools and colleges. *Nesta.* https://media.nesta.org.uk/documents/Future_of_AI_and_education_v5_WEB.pdf

Barus, O. P., Hidayanto, A. N., Handri, E. Y., Sensuse, D. I., & Yaiprasert, C. (2025). Shaping generative AI governance in higher education: Insights from student perception. *International Journal of Educational Research Open, 8*, Article 100452, https://doi.org/10.1016/j.ijedro.2025.100452

Bernabei, M., Colabianchi, S., Falegnami, A., Costantino, F. (2023). Students' use of large language models in engineering education: A case study on technology acceptance, perceptions, efficacy, and detection chances. *Computers and Education: Artificial Intelligence, 5*, Article 100172. https://doi.org/10.1016/j.caeai.2023.100172

Bin-Nashwan, S. A., Sadallah, M., & Bouteraa, M. (2023). Use of ChatGPT in academia: Academic integrity hangs in the balance. *Technology in Society, 75*, Article 102370.

Burger, B., Kanbach, D. K., Kraus, S., Breier, M., & Corvello, V. (2023). On the use of AI-basedtools like ChatGPT to support management research. *European Journal of Innovation Management, 26*(7), 233–241. https://doi.org/10.1108/EJIM-02-2023-0156

Butson, R., & Spronken-Smith, R. (2024). AI and its implications for research in higher education: A critical dialogue. *Higher Education Research & Development, 43*(3), 563–577. https://doi.org/10.1080/07294360.2023.2280200

Celik, I.; Dindar, M.; Muukkonen, H.; Järvelä, S. (2022). The promises and challenges of artificial intelligence for teachers: a systematic review of research. *TechTrends, 66*, 616–630. https://doi.org/10.1007/s11528-022-00715-y.

Chan, C. K. Y. (2023). A comprehensive AI policy education framework for university teaching and learning. *International Journal of Educational Technology in Higher Education, 20*, 38. https://doi.org/10.1186/s41239-023-00408-3

Cotton, D. R. E., Cotton, P. A., & Shipway, J. R. (2023). Chatting and cheating? Exploring the potential impact of ChatGPT on academic integrity. *International Journal for Educational Integrity, 19*(1), 1–13. https://doi.org/10.1007/s40979-023-00119-9

Dong, L., Tang, X., & Wang, X. (2025). Examining the effect of artificial intelligence in relation to students' academic achievement: A meta-analysis. *Computers and Education: Artificial Intelligence, 8*, Article 100400. https://doi.org/10.1016/j.caeai.2025.100400

Eden, C. A., Chisom, O. N., & Adeniyi, I. S. (2024). Integrating AI in education: Opportunities, challenges, and ethical considerations. *Magna Scientia Advanced Research and Reviews, 10*(2), 006–013. https://doi.org/10.30574/msarr.2024.10.2.0039

Funa, A. A., & Gabay, R. A. E. (2025). Policy guidelines and recommendations on AI use in teaching and learning: A meta-synthesis study. *Social Sciences and Humanities Open, 11*, Article 101221, https://doi.org/10.1016/j.ssaho.2024.101221

Godwin, A., Scott, T. D., Potvin, G., Sonnert, G., & Sadler, P. M. (2015). The academic performance index: Creating a more robust and less biased measure of student academic performance. *IEEE Frontiers in Education Conference (FIE)* (pp. 1–8), USA. IEEE. https://doi.org/10.1109/FIE.2015.7344282

Godwin-Jones, R. (2022). Partnering with AI: Intelligent writing assistance and instructed language learning. *Language Learning & Technology, 26*(2), 5–24. 10125/73474

Huong, L. P. H, & Hung, B. P. (2021). Mediation of digital tools in English learning. *LEARN Journal*, 14(2), 512–528. https://so04.tci-thaijo.org/index.php/LEARN/article/view/253278

Johnson, D. L., & Davis, C. G. (2024). Bridging the gap for underserved populations: Personalized AI solutions for college access and learning support. *New Directions for Higher Education, 207*, 47–62. https://doi.org/10.1002/he.20511

Kafa, A. (2025). Exploring integration aspects of school leadership in the context of digitalization and artificial intelligence. *International Journal of Educational Management, 39*(8), 98–115. https://doi.org/10.1108/IJEM-11-2024-0703

Kangwa, D., Msafiri, M. M., & Fute, A. (2025). Exploring the factors that promote a balance between academic integrity and the effective use of GenAI tools in higher education: A systematic review. *Journal of Computer Assisted Learning, 41*(5), Article e70109. https://doi.org/10.1111/jcal.70109

Kasneci, E., Sessler, K., Küchemann, S., Bannert, M., Dementieva, D., Fischer, F., Gasser, U., Groh, G., Günnemann, S., Hüllermeier, E., Krusche, S., Kutyniok, G., Michaeli, T., Nerdel, C., Pfeffer, J., Poquet, O., Sailer, M., Schmidt, A., Seidel, T., & Stadler, M. (2023). ChatGPT for good? On opportunities and challenges of large language models for education. *Learning and Individual Differences, 103*, Article 102274. https://doi.org/10.1016/j.lindif.2023.102274

Kelly, S., Kaye, S., & Oviedo-Trespalacios, O. (2023). What factors contribute to the acceptance of artificial intelligence? A systematic review. *Telematics and Informatics, 77*, Article 101925. https://doi.org/10.1016/j.tele.2022.101925

Khatun, M., Islam, R., Kumar, S., Hossain, R., & Mani, L. (2024). The impact of artificial intelligence on educational transformation: Trends and future directions. *Journal of Information Systems and Informatics, 6*(4), 2347–2373.

Kohnke, L. (2025). Generative AI. In L. McCallum & D. Tafazoli (Eds.), *The Palgrave Encyclopedia of computer-assisted language learning*. Palgrave Macmillan. https://doi.org/10.1007/978-3-031-51447-0_100-1

Kousha, K., & Thelwall, M. A. (2023). Artificial intelligence to support publishing and peer review: A summary and review. *Learned Publishing, 37*(1), 4–12.

Machkour, B., & Abriane, A. (2025). The rise of artificial intelligence in educational management: A prospective analysis on the role of the virtual educational director. *Procedia Computer Science, 257*, 1233–1238. https://doi.org/10.1016/j.procs.2025.03.165

Marín, Y. R., Caro, O. C., Rituay, A. M. C., Llanos, K. A. G., Perez, D. T., Bardales, E. S., Tuesta, J. N. A., & Santos, R. C. (2025). Ethical challenges associated with the use of artificial intelligence in university education. *Journal Academic Ethics, 23*, 2443–2467. https://doi.org/10.1007/s10805-025-09660-w

Nadim, M. A., & Fucio, R. D. (2025). Unveiling the potential: Artificial intelligence's negative impact on teaching and research considering ethics in higher education. *European Journal of Education Research, Development, and Policy, 60*(1), Article e12929. https://doi.org/10.1111/ejed.12929

Nguyen, A., Ngo, H. N., Hong, Y., Dang, B., & Nguyen, B. T. (2023). Ethical principles for artificial intelligence in education. *Education and Information Technology, 28*, 4221–4241. https://doi.org/10.1007/s10639-022-11316-w

Pal, S. (2023). A paradigm shift in research: Exploring the intersection of artificial intelligence and research methodology. *International Journal of Innovative Research in Engineering & Multidisciplinary Physical Sciences, 11*(3). https://doi.org/10.37082/IJIRMPS.v11.i3.230125

Pepin, B., Buchholtz, N., & Salinas-Hernández, U. (2025). A scoping survey of ChatGPT in mathematics education. *Digital Experiences in Mathematics Education, 11*, 9–41. https://doi.org/10.1007/s40751-025-00172-1

Perkins, M. (2023). Academic integrity considerations of AI large language models in the post-pandemic era: ChatGPT and beyond. *Journal of University Teaching and Learning Practice, 20*(2). https://doi.org/10.53761/1.20.02.07

Pikhart, M. & Al-Obaydi, L. H. (2025). Reporting the potential risk of using AI in higher education: Subjective perspectives of educators. *Computers in Human Behavior Reports, 18*, Article 100693. https://doi.org/10.1016/j.chbr.2025.100693

Ramdani, Z. (2018). Construction of academic integrity scale. *International Journal of Research Studies in Psychology, 7*(1), 87–97. https://doi.org/10.5861/ijrsp.2018.3003

Schmidt, T., & Strasser, T. (2022). Artificial intelligence in foreign language learning and teaching: A CALL for intelligent practice. *Anglistik: International Journal of English Studies, 33*(1), 165–184. https://doi.org/10.33675/ANGL/2022/1/14

Shah, Z., Shahzad, M. H., Saleem, S., Taj, I., Amin, S., Almagharbeh, W. T., & Durvesh, S. (2025). Ethical considerations in the use of ai for academic research and scientific discovery: a narrative review. *Insights-Journal of Life and Social Sciences, 3*(2), 183–189. https://orcid.org/0000-0002-8435-1208

Steinmayr, R., Crede, J., McElvany, N., & Wirthwein, L. (2016). Subjective well-being, test anxiety, academic achievement: Testing for reciprocal effects. *Frontiers in Psychology, 6*, Article 1994. https://doi.org/10.3389/fpsyg.2015.01994

Tan, X., Cheng, G., & Ling, M. H. (2025). Artificial intelligence in teaching and teacher professional development: A systematic review. *Computers and Education: Artificial Intelligence, 8*, Article 100355. https://doi.org/10.1016/j.caeai.2024.100355

Tang, K. H. D. (2024). Implications of artificial intelligence for teaching and learning. *Acta Pedagogia Asiana, 3*(2), 65–79. https://doi.org/10.53623/apga.v3i2.404

Temper, M., Tjoa, S., & David, L. (2025). Higher education act for AI (HEAT-AI): a framework to regulate the usage of AI in higher education institutions. *Frontiers in Education, 10*, Article 1505370. https://doi.org/10.3389/feduc.2025.1505370

TEQSA (2022). *What is academic integrity*? https://www.teqsa.gov.au/students/understanding-academic-integrity/what-academic-integrity

Tour, E., Pegrum, M., & Macdonald, S. (2025). AI literacy practices. In L. McCallum & D. Tafazoli (Eds.), *The Palgrave Encyclopedia of computer-assisted language learning*. Palgrave Macmillan. https://doi.org/10.1007/978-3-031-51447-0_93-1

Turková, K., Krásničan, V., Prázová, I., Turčínek, P., & Foltýnek, T. (2025). Adapting to the future: the use of AI tools and applications in university education and a call for transparent rules and guidelines. *International Journal of Educational Integrity, 21*, Article 29. https://doi.org/10.1007/s40979-025-00203-9

Wang, X., Xu, X., Zhang, Y., Hao, S., & Jie, W. (2024). Exploring the impact of artificial intelligence application in personalized learning environments: Thematic analysis of undergraduates' perceptions in China. *Humanities and Social Sciences Communications, 11*, 1644. https://doi.org/10.1057/s41599-024-04168-x

Weber-Wulff, D., Anohina-Naumeca, A., Bjelobaba, S., Foltýnek, T., Guerrero-Dib, J., Popoola, O., Šigut, P., & Waddington, L. (2023). Testing of detection tools for AI-generated text. *International Journal for Educational Integrity, 19*(1), 26. https://doi.org/10.1007/s40979-023-00146-z

Weichert, J., & Dimobi, C. (2024). DUPE: Detection undermining via prompt engineering for deepfake text. arXiv:2404.11408. https://doi.org/10.48550/arXiv.2404.11408

Zhang, K., & Aslan, A. B. (2021). AI technologies for education: Recent research & future directions. *Computers and Education: Artificial Intelligence, 2*, 100025. https://doi.org/10.1016/j.caeai.2021.100025

**Le Pham Hoai Huong** is an associate professor in the English Faculty of Hue University of Foreign Languages and International Studies. Her research interests include sociocultural theory, tool-mediated language learning, and professional development.

**Thuy Thanh Dao** is a Ph.D holder. She works at the Faculty of Social Sciences, Hong Duc University, Thanh Hoa Province, Vietnam. She earned her doctoral degree in Vietnamese Studies and has published numerous research articles in national journals and several papers indexed in Scopus. Her research focuses on socio-economic and cultural transformations in the context of urbanization, as well as culture, heritage, and sustainable tourism development.

MIX
Papier aus verantwortungsvollen Quellen
Paper from responsible sources
FSC® C105338

If you have any concerns about our products,
you can contact us on
**ProductSafety@springernature.com**

In case Publisher is established outside the EU,
the EU authorized representative is:
**Springer Nature Customer Service Center GmbH**
**Europaplatz 3, 69115 Heidelberg, Germany**

Printed by Libri Plureos GmbH
in Hamburg, Germany